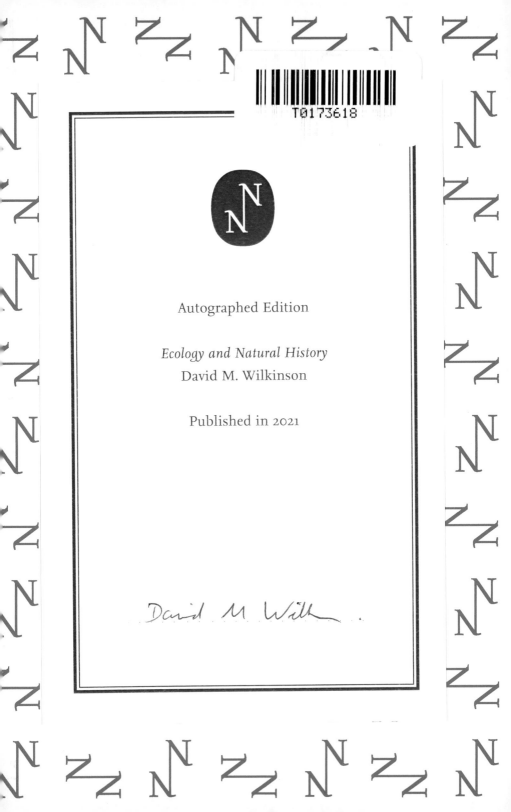

Autographed Edition

Ecology and Natural History
David M. Wilkinson

Published in 2021

THE NEW NATURALIST LIBRARY

A SURVEY OF BRITISH NATURAL HISTORY

ECOLOGY AND NATURAL HISTORY

EDITORS
SARAH A. CORBET, ScD
DAVID STREETER, MBE, FRSB
JIM FLEGG, OBE, FIHort
Prof. JONATHAN SILVERTOWN
Prof. BRIAN SHORT

*

The aim of this series is to interest the general
reader in the wildlife of Britain by recapturing
the enquiring spirit of the old naturalists.
The editors believe that the natural pride of
the British public in the native flora and fauna,
to which must be added concern for their
conservation, is best fostered by maintaining
a high standard of accuracy combined with
clarity of exposition in presenting the results
of modern scientific research.

ECOLOGY AND NATURAL HISTORY

DAVID M. WILKINSON

WILLIAM
COLLINS

This edition published in 2021 by William Collins,
an imprint of HarperCollins*Publishers*

HarperCollins*Publishers*
1 London Bridge Street
London SE1 9GF

WilliamCollinsBooks.com

HarperCollins*Publishers*
1st Floor, Watermarque Building, Ringsend Road
Dublin 4, Ireland

First published 2021

A CIP catalogue record for this book is available
from the British Library.

Set in FF Nexus

Edited and designed by
D & N Publishing
Baydon, Wiltshire

Printed in Bosnia and Herzegovina by GPS Group

Hardback
ISBN 978-0-00-829363-5

Paperback
ISBN 978-0-00-829365-9

Contents

Editors' Preface

This is the 143rd volume in the New Naturalist Library, every one of which has been in some degree ecological, and yet this is the first book in the series to take ecology itself as its subject matter. As the author says in his foreword, the book was in part inspired by the late Sam (R. J.) Berry's *Inheritance and Natural History* (New Naturalist 61, 1977), which used examples from the British Isles to illustrate the science of genetics and evolution. Here, in this present volume of the series, Professor Dave Wilkinson uses British habitats to explicate the science of ecology. One could of course just as easily draw ecological and evolutionary object lessons from anywhere on the planet, and yet where else than in Britain can one reference and explore the very entangled bank that Charles Darwin wrote about in the *Origin of Species*, and see the very orchids that appear in his book on how these flowers are pollinated?

Darwin's Downe Bank is just one of the many ecological locations that you will visit in this book in the company of Professor Wilkinson. His evocative prose, ecological expertise and wonderful photographs combine to produce a fascinating tour that takes the reader simultaneously through the metaphorical and the actual landscapes of ecology. The tour begins at Downe Bank because here we see the ecological complexity that Darwin so clearly understood, and that provides the ecological theatre in which evolution performs its play. Each location visited in this book serves three functions. First, as at Downe Bank, there is its historical significance in the development of science; second, all are habitats chosen because they help us understand a facet of ecology particularly well (interactions among species in the case of Downe Bank); and third, every location has been studied for many decades, providing a clear record of ecological change, which is key to both understanding and conserving our environment. Many of the species that Darwin knew are still present in the environs of Downe – although, as Dave Wilkinson observes, today 'the parakeets would have surprised him.'

Ecology is the science of the relationships between organisms and their environment, including other organisms. It is in extreme environments such as at Cwm Idwal in Snowdonia (Chapter 2) that the physical constraints of temperature and soil become clearly visible, for example in the distribution of arctic–alpine plants. Next stop is Wytham Woods on the outskirts of Oxford, where the ecology of the woodland community has been intensively studied for three-quarters of a century. Many volumes in the New Naturalist Library would be the skinnier but for these studies at Wytham. The focus here is on what the wood tells us about the carbon cycle.

At Moor House National Nature Reserve, we learn of the large significance of the very small. Decades of study at this upland site inform our knowledge of soil invertebrates and microbes. The concept of ecosystems is developed at Windermere, competition between species is revealed on the shores of the Isle of Cumbrae, and cooperation is uncovered in the Cairngorms. Gilbert White's observations at Selborne in Hampshire are the starting point for studying long-term trends in animal populations, but this extends much further afield to the Bass Rock in Scotland and then to a nationwide, nine-decade record of the Grey Heron population in England and Wales. Successional changes in vegetation are seen at Wicken Fen in Cambridgeshire. There is much more yet, but this sketch of the breadth and depth of *Ecology and Natural History* should by now have made you impatient to embark at Chapter 1. Whether you are naturalist, ecologist or no -ist at all, this book will provide pleasure and a new perspective to anyone interested in natural history and its modern scientific scion – ecology.

Author's Foreword and Acknowledgements

'What?' and 'which?' are probably the most common starting questions in natural history. What species was that? Which species are found at this location? However, the obvious follow-on question is 'why?' Why is this particular species found here? Why are some sites more species-rich than others? It's always been ideas – the 'why' questions – that particularly appeal to me, and my aim in this book is to introduce some of the basic ideas of the science of ecology, using examples from British natural history. Ecology is a science where natural history still plays a major role – not just in answering the 'what' questions such as identifying the insects in your sweep-net sample or moth trap, but in generating and answering the 'why' questions too. As Mary Willson and Juan Armesto (2006) put it, 'Very commonly, it is simple natural history observations, planned or unplanned, that tweak the imagination into challenging existing dogma, asking novel questions, and seeing natural phenomena from a different perspective.'

The contingencies of history make Britain a good country in which to base a book of this type. Ecology started to develop here earlier than in many other places, so many classic studies that helped develop the basic ideas of the subject were carried out at British sites. For example, some of the longest-running field experiments in ecology – such as the Park Grass Experiment at Rothamsted (started 1856) or the Godwin Plots at Wicken Fen (started 1927) – are found in Britain. Wytham Woods is arguably one of the best-studied deciduous woodlands in the world, after being acquired by the University of Oxford as a research site in the mid-twentieth century, and there has been almost a century of ecological research based at Windermere in the Lake District, making it one of the most intensively studied lakes around (Fig. 1). Each chapter in this book takes such a classic location as its starting point to introduce a range of ecological ideas – from nutrient cycling, to population ecology, to ecosystems and the workings of the whole Earth system. Although this is a book about the underlying science, the final chapter discusses a number of more applied issues, concerning the effects our species has been having on the wider environment.

Over four decades ago R. J. (Sam) Berry wrote a similar book in this series on genetics called *Inheritance and Natural History*. Sam's idea was similar to mine, to use examples from British natural history to illustrate the basic ideas of his science, in his case genetics; however, the complexities of his subject led to something that looked rather like a textbook (indeed, he said that the book was effectively his final-year lectures, with most of the maths removed). I have tried for something more accessible, while maintaining scientific rigour. So I have no mathematical equations, and I have adopted a policy of using graphs and tables sparingly, mainly illustrating the points with photographs of the species and habitats under discussion – the things you can see in the field. The references should allow the reader in need of more detailed technical background to know where to look. However, this book is similar to Sam Berry's in another way as well. He wrote in his own foreword that, as a reasonably popular account, 'it does not have to be as inclusive and balanced as a textbook' in its choice of topics and examples. The same is true here. For example, neither predation nor parasitism has its own chapter (as would almost certainly be the case in a textbook), although they are covered in several of the chapters discussing other ecological ideas. I have focused mainly on questions in population and ecosystem ecology; a similar book could be written concentrating on ideas in behavioural and evolutionary ecology.

This book has another key theme, namely the diversity of life involved in ecological processes. Indeed, Chapter 4 is entirely devoted to the importance of the microorganisms that make up most of the diversity of life on Earth, when viewed from either a genetic or a biochemical perspective. All naturalists are members of just one species – *Homo sapiens*, ourselves. Think about the implications of this for potential biases in understanding life on Earth. Our taxonomic chauvinism is

FIG 1. Key sites discussed in the book. Sites used as chapter openings are shown in red, and other sites covered at some length (or used in multiple chapters) are shown in blue. At the scale of this map two other Peak District sites (Winnats Pass and the village of Eyam) are in an almost identical location to Ringinglow Bog. Most of these sites are public access – or can be seen from public paths and roads. Access to Wytham Woods is by permit, but these are readily available (a web search should find the current details). The Godwin Plots at Wicken Fen are on a part of the National Trust reserve with a visitor fee for non-members.

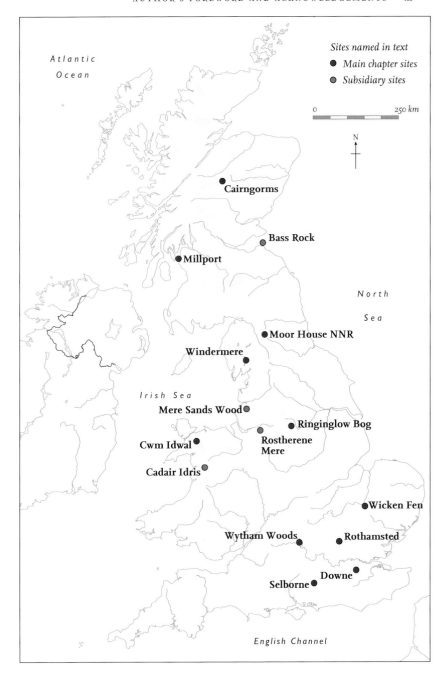

Atlantic Ocean

Sites named in text
● Main chapter sites
◐ Subsidiary sites

0   *250 km*

N

●Cairngorms

◐ Bass Rock

●Millport

North

Sea

●Moor House NNR

Windermere●

Irish Sea

Mere Sands Wood◐

●Ringinglow Bog

Cwm Idwal●

◐ Rostherene Mere

Cadair Idris◐

●Wicken Fen

Wytham Woods●   ●Rothamsted

Downe●

Selborne●

English Channel

easily seen if you look at TV natural history programmes, popular natural history books or even university-level ecology textbooks – they are all preferentially populated with organisms that are rather like us, such as mammals and birds (Kokko 2017). And yet in most ecological processes, microbes, fungi and plants are far more important than birds or mammals.

I have tried to take examples from across the full range of biological diversity, as restricting the range of organisms that we use to think about ecology will necessarily restrict the range of answers we are likely to come up with. For example, we usually partition life into organisms that can photosynthesise and those that eat other organisms (incorrectly equating this split with the terms 'plants' and 'animals'). However, many microorganisms do both, a lifestyle called mixotrophy. Imagine if a mixotrophic amoeba could write an ecology book – would it consider organisms that are so specialised that they can only get their energy by just one method strange exceptions to the more usual way of things? That amoeba is mixotrophic because it has algae living inside its cell, something that seems alien to us but is again common in many organisms, and deep in geological history gave rise to the evolution of plants. In that imaginary book authored by a protozoan, there might well be only one or two illustrations of vertebrates – creatures so far away from what the author considers normal that they are only discussed in passing as interesting oddities.

The fact that all science is carried out by people not only affects how we view biodiversity, but necessarily means that it is 'a socially embedded activity' too (e.g. Gould 1981). Society's norms cannot help but affect the way science and academia function. While this must be the case – scientists are people too – exactly how you interpret this insight has been something of a Pandora's box, and the subject of much academic debate and misunderstanding (Latour 1999). However, this socially embedded insight has implications for a book such as this one which concentrates on introducing the basic ideas – for these were often developed decades ago, at a time when the sex ratio of most areas of science was strongly skewed towards males. In this respect, science was reflecting wider British society at the time. So many of the key studies I write about were carried out by men. This skewed sex ratio is no longer as apparent in academic ecology, and indeed the acknowledgements from this book provide a nice illustration of these changes. Friends and colleagues thanked for providing technical advice on draft chapters or helping with fieldwork form a population that is close to a 50:50 sex ratio – while those, mainly retired, thanked for helping with points of history by supplying memories of when they were young scientists are almost all male. However, even today if you stand before a group of British university students attending an ecology lecture it's still likely that the audience will not match the wider country in its ethnic origins – suggesting that we are still missing out on much potential talent in ecology. This matters for all sorts of reasons, including that we need all the talent we can find. For ecology is a science that underpins our attempts to address many of the most challenging global problems.

ACKNOWLEDGEMENTS

Many people have helped with the process of writing this book. Jonathan Silvertown, Hannah O'Regan and Graeme Ruxton have read and commented on all the chapters. All three have had a significant role in improving the text. Jonathan (my editor as well as a plant ecologist and writer) was very influential in early discussions on how to structure such a book. Hannah is my wife and has provided support and encouragement through the multi-year writing process. Graeme read all the chapters purely out of interest and friendship – his enthusiasm for this book is greatly appreciated. The following commented on chapters in their particular area of expertise: Tom Barker, Sarah Dalrymple, Jane Fisher, Stuart Humphries, Edward Mitchell and other members of the soil ecology group at the University of Neuchâtel, Tom Sherratt and Sandra Varga. Several people provided comments on various points of detail: Richard Betts, Martha Crockett, Stefan Geisen and Becky Yahr. Help and/or advice with fieldwork came from Charles Deeming, Jenny Dunn, Paul Edy, Sarah Elton and Anne-Marie Nuttall. Historical information (recollections of when they were younger) came from Bill Block, John Coulson, Alastair Fitter, Ian Hodkinson, Brian Huntley, John Lawton, Penny Oakland, Donald Pigott and Richard West. It takes many talented people to turn a manuscript into a finished book – including Myles Archibald and Hazel Eriksson at HarperCollins, and Namrita and David Price-Goodfellow of D & N Publishing. Lisa Footit did a great job compiling the index, while Hugh Brazier's contribution merits particular acknowledgement, as it went well beyond the grammatical tweaks normally associated with the role of copy-editor. The photos are my own, except for three by the late Bryan Nelson (thanks to June Nelson for supplying these), two by Becky Yahr, one by Angela Creevy (from our work at Mere Sands Wood nature reserve), and one by my father Lionel Wilkinson – who sadly didn't live to see the completed book. This book is dedicated to Hannah with love, and in memory of Lionel.

CHAPTER 1

The Entangled Bank

It is an afternoon of uncertain weather in early May. The clouds repeatedly thickening, promising rain, then breaking to reveal patches of blue – allowing the sun to briefly warm the ant hills on the grassy slope in front of me. I am sitting on a small fragment of chalk grassland on the side of the Cudham valley in Kent. This remnant of a previously much more diverse countryside is now just a thin sliver of open habitat on an otherwise wooded slope (Fig. 2). The valley bottom below me illustrates the

FIG 2. Downe Bank nature reserve in the Cudham Valley, Kent, in early May 2019. A short walk from Down House, and a favourite afternoon walk for Charles and Emma Darwin.

twentieth-century fate of much of this species-rich vegetation – converted
to bright green fertilised fields, species-poor and dominated by grass. Fields
'improved', in the language of modern agriculture, although that's not how a
plant conservationist would see it.

At first glance the vegetation on the bank looks simple, comprising grassland
with a small path running between ant hills. Take a closer look, however, and
it reveals itself as a complex tangle of mosses, grasses and a wide variety of
herbs. There are nine species of orchid growing here, although this early in the
year only Common Twayblade is flowering (Fig. 3). The orchid-rich character
of this bank is not new – in the 1860s, obsessed by the complexities of orchid
reproduction, Charles Darwin fondly referred to this site as his 'orchis bank'.

FIG 3. Common
Twayblade *Neottia
ovata* flowering on
Downe Bank, one of
nine species of orchids
that grow on Darwin's
orchis bank.

The name fitted the site well in the 1860s when many more British orchid species were placed in the genus *Orchis* than is the case today, and orchis was also used in the English name of several orchid species. The Darwins lived in Downe village only a short walk away, and the orchis bank was the objective of a favourite afternoon walk for Charles and his wife Emma, and possibly one of the inspirations for Darwin's metaphor of an 'entangled bank' that famously appears in the final paragraph of *On the Origin of Species*. Today it's still a quiet stroll from Downe along a footpath beside the Cudham Road to reach the reserve. As a classic description of the complexity of nature that ecologists strive to understand, Darwin's entangled bank still works today. Indeed, this book is an introduction to the ideas that help us start to untangle the entangled bank, and understand the complexities of ecology.

Darwin (1859) closed one of the most famous books in the history of science with this description of the English countryside he knew:

> *It is interesting to contemplate an entangled bank, clothed with many plants of many kinds, with birds singing on the bushes, with various insects flitting about, and with worms crawling through the damp earth, and to reflect that these elaborately constructed forms, so different from each other, and dependent on each other in so complex a manner, have all been produced by laws acting around us.*

Many of the banks and field margins around Downe in the mid-nineteenth century would have matched this description, and while Downe Bank is often cited as the origin of the entangled bank (simplified to a 'tangled bank' in later editions of *The Origin*), Janet Brown (2002), one of Darwin's most important biographers, rightly deploys the caution and scepticism of a good historian, writing that the orchis bank 'may well have been the same sweetly tangled bank'. Whatever its role in inspiring Darwin's classic image of biodiversity, the Darwin connection gives Downe Bank added importance in British natural history, and it was the first reserve acquired by the Kent Wildlife Trust, in 1962.

That day in May Brimstone butterflies *Gonepteryx rhamni* were 'flitting about', yellow against the bright green of the unfolding leaves on the coppice at the base of the bank, while overhead flew Ring-necked Parakeets *Psittacula krameri*. While much has not changed on the bank since Darwin's time, the parakeets would have surprised him. It's a bird he probably never saw in the wild – its native range is central Africa, India and nearby countries (places he never visited on the *Beagle* voyage). There were some escapes from captivity into the wild in Britain, including a population that survived for a few years in Norfolk around the time of Darwin's death, but our current population appears to have originated from multiple escapes of captive birds during the twentieth century (Heald *et al.* 2020). This colourful intrusion of tropical diversity into the Kent countryside illustrates one of the many ways in which humans have changed the ecology of the planet – moving large numbers of organisms to locations far from their original home – a topic returned to in Chapter 12.

WHAT IS ECOLOGY?

Using Darwin's metaphor of an entangled bank to introduce the science of ecology raises a basic question – what do we mean by ecology? The word 'ecology' gets used in multiple ways, several of which are compatible with using Darwin's orchis bank as the opening example in this book. Say 'ecology' to many people and they will often think of nature conservation and attempts to address a whole host of environmental problems – from 'saving the whales' when I was growing up in the 1970s, to fears that the whole planet is potentially now in need of saving. This book doesn't address ecology in this sense (except in the final chapter, which gives a necessarily selective brief overview of some aspects of ecology as applied to environmental problems). Instead it focuses on the science of ecology – especially the ecological concepts that attempt to explain the workings of Darwin's entangled bank. Why is this bank (and the world at large) so diverse, and what are the relationships between the plants of many kinds, the birds singing on the bushes, and the various insects flitting about, not to mention the worms crawling through the damp earth and a host of other small and often overlooked organisms? The importance of microorganisms and fungi in ecological processes is one of the recurring themes in this book. Key ecological processes are often driven by obscure organisms that seldom feature in TV natural history documentaries or books and magazine articles written by, and for, naturalists (Fig. 4).

The term 'ecology' was coined (originally as 'oecologie') by Ernst Haeckel and popularised in a widely read book of 1866 where he defined ecology as 'the whole science of the relations of the organism to the environment' (Egerton 2012). Our modern definitions are still very similar – for example, 'the scientific study of the abundance and distribution of organisms in relation to other organisms and environmental conditions' (Ricklefs & Relyea 2014). Haeckel certainly didn't singlehandedly invent ecology; as so often in the history of science, it is impossible to identify a clear starting date, and many people were starting to do what we would now consider ecological studies long before 1866 (one of many examples was Gilbert White; see Chapter 8). However, giving this area of science

FIG 4. Lichen-covered rock on Cadair Idris, Snowdonia. The large pale green lichen is Map Lichen *Rhizocarpon geographicum* (the grey patches are other species of lichens). The main body of a lichen is made up of fungus, but the 'organism' itself is a composite of fungus and a range of species of microbes (see Chapter 7).

a name and a formal definition obviously helped focus people's attention on this way of thinking about the world, although it was the end of the nineteenth century before ecology started to become a growing area of research.

In Britain, plant ecology developed more rapidly than animal ecology, with Arthur Tansley being a particularly influential figure in the early decades of the twentieth century. It would however be incorrect to assume that the key driver for ecological research has always been the sort of environmental concern that people would now often attribute to scientists working in this area. As the science historian Peter Bowler (1992) has pointed out, many early ecological studies were 'often initiated by scientists who hoped to modify the natural balance in order to allow sustainable exploitation'. This was particularly the case for some of the early ecological researchers in the United States of America, and it is still important – for example in applying ecological ideas to try and formulate a more sustainable agriculture. Whatever the underlying reason for studying ecology, the science is underpinned by a number of basic ideas, and the aim of this book is to introduce some of these using examples drawn from British natural history. The rest of this opening chapter discusses a couple of very basic and fundamental concepts – with many more detailed ideas considered in the subsequent chapters.

ISLANDS OF ORDER IN A SEA OF CHAOS

An obvious question in ecology is this – where does an organism get its energy from? Common answers are that the energy comes from photosynthesis, or that it comes from eating other organisms – be that as a herbivore eating plants (Fig. 5) or as a carnivore eating other animals. As discussed in more detail later in this book, however, other options are available, such as feeding on dead biological material (Fig. 6) or mixing photosynthesis with consuming other organisms.

To understand why the acquisition of energy is such a fundamental process in ecology requires a brief excursion into the ideas embodied in an area of science called thermodynamics. This sounds a long way from anything to do with natural history, but as a science that's all about the movement of energy in the form of heat it's very relevant to many 'why' questions in natural history. The early history of thermodynamics makes this obvious. It was developed as an area of theory applicable to steam engines – how they are fuelled (effectively, what they eat) and how efficiently they can use these sources of power. These are also questions we can ask about the plants, birds and worms in Darwin's entangled bank.

Fundamental to understanding energy use by organisms (or steam engines) is the idea of the second law of thermodynamics. This 'is of central importance

FIG 5. Alder Leaf Beetles *Agelastica alni* feeding on Alder *Alnus glutinosa* leaves at Askham Bog on the outskirts of York in 2019. This beetle was thought to have become extinct in Britain in the 1940s, but it reappeared in 2004, and is now becoming common at some locations in northern England.

FIG 6. Turkeytail *Trametes versicolor* in Bunny Old Wood, Nottinghamshire. A very common fungus growing, and feeding, on dead wood. This is a very variably coloured species, and one of the commonest polypores in Europe (Kibby 2017).

in the whole of science, and hence in our rational understanding of the universe, because it provides a foundation for understanding why *any* change occurs' (Atkins 2007). Peter Atkins, who wrote these words, is a distinguished physical chemist, but the second law is just as central to ecology as it is to chemistry. This should not be a surprise when you consider that metabolism (the processes in cells by which energy is stored or released) is a chemical process. This approach to thinking about thermodynamics and ecology, described below, is a brief summary of a somewhat more technical discussion in my earlier book *Fundamental Processes in Ecology: an Earth Systems Approach* (Wilkinson 2006).

Look in a physics text or popular science book (such as the one by Peter Atkins from which I have just quoted) and you will see that the second law is usually defined in relation to a concept called 'entropy'. One way to think of entropy is that it is a measure of the disorder, or our lack of information, about the state of things – this is a horribly inexact way of describing entropy, but any attempt at anything more rigorous quickly becomes rather technical (Thorne & Blandford 2017). One brief way of stating the second law of thermodynamics is that 'the entropy of the universe is always increasing' – so we should expect things to become more disordered over time. This certainly seems true of my study as I am working on a book, and physics suggests it is true of the universe as a whole. However, as naturalists we can sidestep ideas of entropy and instead use a more intuitive concept called 'free energy'. This is simply the amount of energy available for doing useful work – for example, walking to the bookshelves behind me as I write, to check physics textbooks in an attempt to make sure what I am writing is at least approximately correct (give or take the odd simplification).

Why should free energy matter to an ecologist? Think about the look of an organism such as the orchid in Figure 3 or the beetles (or the Alder leaf) in Figure 5, or look at yourself in a mirror – all these examples look ordered, and in many cases they are bilaterally symmetrical too (i.e. one side is a mirror image of the other). In a universe dominated by the second law, how do such organisms manage to stay so organised rather than falling apart? In a neat phrase, Lynn Margulis and Dorion Sagan (1995) encapsulated this idea as organisms representing 'islands of order in an ocean of chaos'. The way organisms maintain this order – and temporarily save themselves from death and the ocean of chaos – is to use energy from their environment to maintain their ordered structure. As no organism, or machine, can be 100 per cent efficient (this follows from thermodynamics too), then all organisms must also be producing some waste. The ecological implications of these fundamental ideas from physics can be summarised as:

energy $\rightarrow$ organism $\rightarrow$ waste

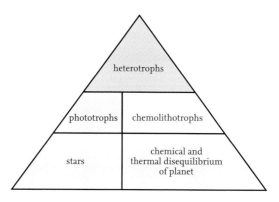

FIG 7. The pyramid of free energy production (after Lineweaver & Egan 2008). Traditionally in ecology a pyramid – with photosynthetic organisms at the base, a smaller number of herbivores above, and an even smaller number of carnivores above them – is used to show the amount of energy available for use as one moves along food chains. This figure is effectively a more fundamental version of the same idea – applicable to any planet with life. Energy can come either from the planet's star (in our case the sun) or from chemical sources. The light can be used as a source of energy by photosynthetic organisms (phototrophs), while chemical sources are used by chemolithotrophs. These organisms can then provide food (and energy) for heterotrophs. See Chapter 3 for a more detailed discussion of food chains and energy sources.

So all living things are using their environment as a source of energy (and other resources) while also using their environment as a dumping ground for waste products; physics gives them no other option. Much of ecology, from food chains to nutrient cycling, involves sorting out the detailed implications of this very simple idea. As described in more detail in later chapters, the source of energy can be sunlight (for photosynthetic organisms), chemical sources of energy, or obtaining energy by eating other organisms (Fig. 7). All of these sources of energy are being utilised on Darwin's orchis bank – and indeed in the majority of ecological systems. There are occasional apparent exceptions to this rule: for example, if you are a caving naturalist then you will be studying systems where there is no light, and so no photosynthesis – although products of photosynthesis may still have entered the cave from sunlit systems outside (Fig. 8).

FIG 8. Herald moths *Scoliopteryx libatrix* hibernating in a cave in southern Cumbria. By entering the cave they provide a potential food source for cave-living organisms based on photosynthesis outside the cave – because their caterpillars feed on a number of different tree species.

TRADE-OFFS – A KEY ECOLOGICAL CONCEPT

Leaving Darwin's bank, if you walk back along Cudham Road to the village of Downe you will find Darwin's former home, Down House, which is now a museum. In Darwin's day the gardens were a mix of flower gardens, kitchen gardens, multi-compartmented greenhouse, along with orchard, meadow and small fragments of woodland. These gardens provided not only outdoor space for the family and a source of fresh vegetables for the kitchen, but also a place where Darwin could walk and think, and a site for his experiments. The wooden superstructure of the greenhouse at Down House was replaced around 1900, but the basic layout and the brick walls are as used by Darwin, who had wide-ranging botanical interests, one of which was carnivorous plants. Today, as in Darwin's time, part of the greenhouse is filled with carnivorous plants from around the world (Fig. 9). These plants form a charismatic introduction to ideas of trade-offs in ecology and evolutionary biology, as their leaves have to perform two rather different functions – namely photosynthesis and trapping their prey. By a trade-off we mean 'any situation in which the quality of one thing must be decreased for another to increase', although the plant ecologist Peter Grubb (2016), who favours

FIG 9. The greenhouse at Down House, showing a range of carnivorous plants, including tall tropical pitcher plants and smaller sundews. Although sundews grew in the wild near Down House, Darwin brought them into the greenhouse to make his experiments easier to carry out.

this definition, does note that sometimes 'trade-off' has been used in ecology in a looser and less rigorous way. Before discussing this in more detail I will use a non-natural history example to introduce the key idea of trade-offs, using sports science to introduce the basic idea before applying it to carnivorous plants.

Anyone who has ever taken even a passing interest in sport will be familiar with the idea that different types of physique tend to be associated with success in different sports. If you were confronted with three high-standard sports people and told that one was a basketball player, another a jockey and the third a sumo wrestler, you would have little difficulty in guessing which one was which. The fact that different sports are associated with different physiques means that there is no one body type that can excel in all sports. A sport that requires substantial physical bulk tends to rule out also excelling as a jockey (think of the poor horse!) or as a rock climber, where too much weight is a disadvantage when fighting gravity on overhanging rock (Fig. 10). One way to make such arguments

FIG 10. Nigel Smart climbing the route 'Cave Eliminate' on Stanage Edge in the Peak District. The main work done by a rock climber is to move his or her body weight upwards, against gravity. As muscle is a heavy tissue, what really matters is the climber's power-to-weight ratio, and there is a trade-off between amount of muscle and body weight – the more muscle you have the more weight you need to move upwards against the force of gravity, but without at least some muscle you will make no progress at all.

more quantitative – a favoured approach of most scientists – is to use data on the performance of top-class decathletes, where they have to compete in ten different events in order to win. Raoul Van Damme and colleagues (2002) analysed the performance of 600 'world class decathletes' and found nice illustrations of the importance of trade-offs. For example, athletes who did really well at the distance running (1,500 m) tended to do less well in the shot put, and vice versa. This is unsurprising, since if asked to picture a typical distance runner most people will visualise a much thinner athlete than if asked to picture a typical shot-put specialist. The decathlon is interesting as it tends to select for athletes who are rather good in all the events – that's how you win – but may not necessarily be top in any particular event. The key point for thinking about natural history is the authors' conclusion that 'in an environment in which the selection criterion is combined high performance across multiple tasks, increased performance in one function may impede performance in others'. That is effectively Peter Grubb's preferred definition of trade-off, as given above.

With this conclusion in mind we can return to thinking about the carnivorous plants that so interested Darwin. His fascination with such plants grew from simple natural history observation:

During the summer of 1860, I was surprised by finding how large a number of insects were caught by the leaves of the common sun-dew (Drosera rotundifolia) *on a heath in Sussex. I had heard that insects were thus caught, but knew nothing further on the subject.* (Darwin 1875)

The sundew (the usual English name for this species is now Round-leaved Sundew), along with other species of insectivorous plants, quickly made its way into Darwin's greenhouse, and he started a long series of experiments so that he came to know substantially more on the subject. My own experience suggests that 160 years later this is still a good way of doing biology, seeing something in the field, or down a microscope, and thinking 'I wonder how that works?' Alternatively, sometimes it can be reading about something and asking 'how does that work?' or 'can that explanation really be correct?' Darwin realised that many of these plants grow in wet boggy conditions and that because of this they may struggle to get all the nutrients they need from the soil. Their answer is to get these nutrients, and possibly some energy too, from catching and digesting small invertebrates. This presents a problem, because being green plants they need to photosynthesise, and yet their insect traps are often modified leaves. And a leaf that makes a good solar panel doesn't make a good trap, so there have to be trade-offs.

The typical habitat for carnivorous plants is brightly lit, but wet and with low soil nutrients. The suggestion is that there is a trade-off between light collected for photosynthesis and ability to trap insects – ideas first proposed by Thomas Givnish and colleagues during the 1980s and subsequently developed by others (Ellison & Gotelli 2009). Obviously a leaf should be a nice flat surface to intercept lots of light, but to trap insects it has to be modified – for example, the leaves of a butterwort are curved to reduce the chances of small insects escaping the sticky leaves (Figs. 11 & 12). The fact that such plants tend to live in places with lots of light presumably makes the cost of having leaves that are less efficient at trapping light less of a problem.

Another potential trade-off is that between attracting insects as prey and attracting them as pollinators. Many insectivorous plants have their flowers on long stems (Fig. 11), which may be a way of minimising this problem, and/or

FIG 11. Common butterwort *Pinguicula vulgaris* growing in the open – well lit – on a wet streambank in the Yorkshire Dales. This is a typical habitat for this plant, as Darwin (1875) noted – 'This plant grows in moist places, generally on mountains.' As described by Darwin, the 'leaves are deeply concave, and project upwards' (see Fig. 12). Note also the flower on a long stalk, which keeps it away from the leaves where insects are trapped.

FIG 12. Close-up of butterwort leaves, showing the remains of trapped insects. The glandular surface of the leaf can be seen, which is sticky and releases digestive enzymes. Also note the overhang made by the edge of the leaf, which makes it more difficult for trapped insects to escape the sticky leaf surface.

a way of making their flowers easily seen by potential pollinators – indeed, experiments on two species of South African sundews found no evidence that pollinating insects were trapped, but did find fewer pollinator visits to plants with artificially shortened flower stalks (Anderson 2010). Also it may often be the case that the insect species trapped by the plant for food are much smaller than those used as pollinators.

An additional example of potential trade-offs is seen in the traps of Greater Bladderwort *Utricularia vulgaris* (Friday 1991). This aquatic plant has underwater traps which catch small invertebrates such as tiny crustaceans. These traps range in length between 1 and 5 mm, and the cost in resources to the plant makes it as expensive to grow one large trap as it would be to grow around 40 small ones. So there is a potential trade-off between number of traps (the more you have the more likely you are to catch something) and trap size (bigger traps can catch larger and more valuable prey). Similar arguments can be made about seed or egg size. Should you opt for lots of small ones or a few large ones?

One of the reasons why the idea of trade-offs is important, and will appear in several of the chapters in this book, can still be seen in Darwin's greenhouse if you visit Down House today (Fig. 9). There is not just one species of carnivorous plant on display but many – and globally there are over 600 species, probably quite a few more depending exactly on your definition of carnivorous plant (Ellison & Gotelli 2009). Why there is more than one species is a question that applies not just to carnivorous plants but to any group of organisms. Indeed, you can ask it of all life: why not just have one species? Why all this confusing diversity in Darwin's entangled bank?

One answer (there are others – see for example Chapter 10 on ecological niches) is that trade-offs mean there is no such thing as the perfect organism, for it is inevitable that improving its ability to do one thing will make it less good at doing others. So rather than having one jack-of-all-trades species, Darwin's natural selection has given us a diversity, all with different adaptations. This can be seen with common garden birds: a Woodpigeon *Columba palumbus* is far too heavy to feed like a Blue Tit *Cyanistes caeruleus* by hanging on thin branches, but its greater size allows it to process types of plant material, such as leaves, that are unavailable to a small bird with a necessarily limited gut size. And neither of these species is able to access soil invertebrates in the way thrushes can. Because it's hard to evolve a bird that is equally adept at eating worms, aphids and leaves, we have a diversity of garden bird species to watch. Just as, to return to the sports science example used above, we have a diversity of sports people, from lean-looking distance runners to more bulky heavyweight boxers.

CONCLUDING REMARKS

Ecology is the science of understanding the multiple relationships between organisms and their wider environment, and this book describes many of the important ideas of ecology using examples drawn from British natural history. This chapter introduced two key concepts relevant to the understanding of ecology; many more detailed ideas are discussed in the following chapters. Fundamental to life is its battle against the second law of thermodynamics, which means that organisms have to draw energy, and other resources, from their environment in order to live. In addition, thermodynamics tells us that all organisms must be producing some waste, as nothing can be 100 per cent efficient in its use of energy. These ideas form the starting point for discussions of energy flow and nutrient cycling within ecosystems, discussed in later chapters. Over the last few decades, ideas of trade-offs have become increasingly important in ecology. When I was a student in the 1980s they were seldom given much emphasis, but today they are widely applied – arguably sometimes over-applied – in ecology. A trade-off can be defined as any situation in which the quality of one thing must be decreased for another to increase, and it is a concept that will be referred to – both directly and implicitly – in several of the chapters in this book.

CHAPTER 2

Cwm Idwal and the Nature of the Environment

Winter, early morning, and I am sitting on the side of a moraine overlooking the lake with the crags of the Devil's Kitchen cliffs rising behind me. For once I am alone in this often busy cirque – even binoculars don't reveal any people in view. The only sounds come from the wind, the snowmelt water in the mountain streams, and a family of Whooper Swans *Cygnus cygnus* on Llyn Idwal below me. For perhaps an hour I watch the distant swans and take photographs, with the light playing on the snowy mountainsides as the clouds shift in the wind. Then, when walkers start to appear in the distance, I pack up and head higher, chasing the retreating solitude up the mountain.

The potential harshness of mountain landscapes, like that winter morning in Cwm Idwal in Snowdonia, allows you to see a pared-down natural history – where key interactions between organisms and the environment may be more easily observed. As W. H. Pearsall wrote in 1950, in his classic New Naturalist *Mountains and Moorlands*, 'to the biologist at least, highland Britain is of surpassing interest because in it there is shown the dependence of organism upon environment on a large scale'. Following Pearsall's lead, this chapter uses the uplands of Britain as the setting for an introduction to some basic ideas on the importance of the environment in understanding natural history.

To have Cwm Idwal (Fig. 13) so empty of people was a rare treat. It is easily accessible from the nearby road and is one of the classic sites in British physical geography. Field trips of school or university students are not unusual, albeit scarce in the very depths of winter when the weather can be arctic. It is also a classic British site for arctic–alpine plants – perhaps the best such site south of

FIG 13. The view down into Cwm Idwal from part way up the Devil's Kitchen cliffs. Llyn Idwal fills the bottom of the Cwm, with the more distant lake being Llyn Ogwen.

the Scottish border. Indeed, the Cwm has been renowned for these plants since the early botanical explorations of the seventeenth century (a time when botanists could still run foul of highwaymen – you would have to travel further afield for that today). Like several other Snowdonian crags, the cliffs of Cwm Idwal went on to suffer from excessive collecting – especially in the nineteenth century, when some of the rare plants had considerable commercial value if sold to wealthy collectors (Marren 1999). It's a story of depletion, rarity, and occasional local extinction that we usually associate with birds' eggs and butterflies, but the toll on many rare plants was also significant.

THE PHYSICS AND CHEMISTRY OF THE ENVIRONMENT

Why is Cwm Idwal such a good place for arctic–alpine plants of limited distribution? Plants here include Roseroot *Rhodiola rosea*, Purple Saxifrage *Saxifraga oppositifolia*, Starry Saxifrage *Micranthes stellaris* (Figs. 14 & 15), Green Spleenwort *Asplenium viride*, and Alpine Clubmoss *Diphasiastrum alpinum*, along with the real rarities such as Oblong Woodsia *Woodsia ilvensis* and Tufted Saxifrage *S. cespitosa*. In fact this is just a special case of a more general question,

FIG 14. Purple Saxifrage *Saxifraga oppositifolia* flowering in Cwm Idwal, with the Devil's Kitchen cliffs in the background.

namely what is it about the environment at a site that makes it suitable for one species but not another? This question is the central theme of this chapter, and can also be a key question in conservation when thinking about how to manage the habitat of a rare species.

For the Cwm Idwal plants, part of the answer is in their description as 'arctic–alpines', with its obvious connection to the climate on the mountainside. The weather can indeed seem challenging, at least to a naturalist, and even more so to some of those students on field trips if their outdoor clothing turns out to be inadequate. In a recent book on British mountain plants Michael Scott (2016) points out that Cwm Idwal is mentioned in at least three previous volumes in the New Naturalist series, and that each of the authors describes poor weather when plant hunting in the Cwm – something Scott too experienced on his visit to

FIG 15. Starry Saxifrage *Micranthes stellaris* in Cwm Idwal – characteristically, this individual was growing in a damp, shaded location amongst the boulders at the back of the Cwm. This plant has usually been considered to be a member of the genus *Saxifraga*, but molecular evidence has shown that it deserves to be placed in the separate genus *Micranthes* (Stace 2019).

the site. Bruce Campbell was one of these 'New Naturalists', and was particularly impressed by the weather, writing that his visit

> *coincided with some of the most atrocious summer weather in which natural history field-work could be attempted … [it] was the worst day on which I have ever been outdoors in my life.* (North et al. 1949)

However, Campbell was perhaps over-selling the weather's awfulness a little; his range of experience of mountain weather may have been a bit limited, as he was not a mountaineer, but primarily an ornithologist whose most important research was on woodland birds (briefly described in Chapter 10). The weather in the Cwm is not unremittingly bad, of course. I must have visited well over 50 times over the years (and writing this chapter has provided the excuse for several more visits), and I have indeed experienced cold rainy days of the sort that are common in Snowdonia, as well as freezing conditions in which huge icicles form on the Devil's Kitchen cliffs to create the 'Devil's Appendix', one of the classic ice climbs of north Wales – but visits have also included summer days where it seemed far too hot as I slogged uphill out of the Cwm towards the heights. On days like that, a plant (or naturalist) is in more danger of dehydration than freezing, but overall it is indeed often cold and wet in the Cwm.

The icicles and ice climbs of an increasingly rare cold winter are just a shadow of the Cwm's past history. The whole architecture of the place shouts ice. That's what many of those parties of geography students are supposed to be looking at; the moraines, the glacially smoothed rocks and the shape of the cirque itself are the telltale signs of the ice sheets and glaciers of the last glaciation, at least once you know how to read them.

Darwin came here too, visiting first in 1831 and then again in 1842 after returning from his round-the-world voyage – this was also the year he moved to Down House with its nearby entangled bank. He described the Cwm as a 'wild amphitheatre' (Darwin 1842), but as he recalled in the 'autobiographical sketch' written towards the end of his life, on his first visit he saw no evidence of past glaciers. However, on his return visit – with his mind now primed with the new and controversial idea that Britain had experienced a past ice age – he saw evidence all around him so that, as he wrote, 'a house burnt down by fire did not tell its story more plainly than did this valley' (de Beer 1974). This quotation now appears, along with a portrait of Darwin and illustrations of some of the Cwm's plants, on a slate slab by the start of the main path to Cwm Idwal at Ogwen Cottage. It's a nice illustration that science isn't merely a matter of looking at the evidence and using this to reach a logically obvious solution; you need theory to guide you in what you are looking at before you can see things that to later generations will seem almost trivially obvious.

The climate of Cwm Idwal today may no longer be glacial, but it's certainly tough enough to limit which plants can grow here, so limiting the competition from other plants against the arctic–alpines. But many mountains are similar in that respect. The other important aspect at Cwm Idwal is geological – not the geologically recent glacial land forms but the chemistry of the rocks themselves. Some of the rocks here are unusually lime-rich by the standards of the volcanic rocks of Snowdonia (Scott 2016). This means that there are more nutrients available to the plants than is the case at many sites in the British mountains (which tend to be built of hard rocks that usually release any nutrients very slowly), so favouring plant growth on the crags of the Cwm. However, the climatic conditions make it hard for faster-growing nutrient-loving lowland plants to outcompete the various arctic–alpines that attract the botanists to the site, so allowing these plants to survive here.

Such aspects of the physics and chemistry of the environment (such as a cold climate, or nutrients from the weathering of rocks) have long been recognised as important in helping to determine what species can survive at a particular site. In the 1830s Charles Lyell, Darwin's friend and mentor, used many examples of the effects of climate on the distribution of organisms in his influential three-volume *Principles of Geology* (Wilkinson 2002). This approach to natural history had already been developed in some detail during the late eighteenth century, especially by Alexander von Humboldt – whose thinking was greatly influenced by the distribution of mountain plants (Wulf 2015).

Introductory textbooks of ecology often refer to these physical and chemical factors as 'abiotic' – that is, they are assumed to affect life, but not to be determined by life. In the early twentieth century Victor Shelford, at the University of Chicago, suggested what he called the 'law of toleration' to describe the effects of such abiotic factors on a species' distribution. He suggested that for a factor such as summer temperature, or the concentration of a particular nutrient in the soil, there was for any organism an optimum value, along with maximum and minimum values beyond which that species could not survive. Although Shelford (1931) originally applied these ideas slightly differently, today this optimum is usually illustrated for a single factor – such as soil moisture – with the distribution shown by a 'bell-shaped' curve (more technically a 'normal' curve, Fig. 16). This curve has the great advantage of being both visually simple and a mathematically well-understood distribution. It is often used, for example, in attempts to model how a plant or animal may respond to climate change. However, its simplicity is obviously a bit of a caricature of the complexity of the natural history of real species. As the mathematician and science writer Ian Stewart (2012) points out in a nice accessible essay on this curve, its great advantage is that it can be used to model a wide range of real-world examples, from games of chance to agricultural crop trials, but there is also 'an unfortunate

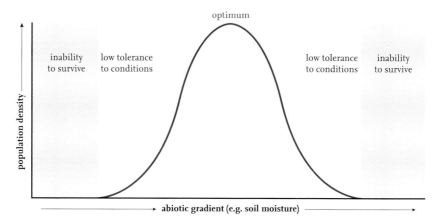

FIG 16. The textbook representation of the idea of limits of tolerance along an abiotic gradient. Population sizes peak at optimal conditions, and as one moves away from the optimum in either direction the size of the population that can be supported by the environment declines until a point is reached where it is unable to survive.

tendency to default to the bell curve as if nothing else existed'. Theoretical ideas in natural history, or in any other area of science, regularly have to try and tread a thin line between being simple enough to be useful and widely applicable, while not oversimplifying to the point of being useless when applied to the complexities of what we see in the mountains or forests of the real world.

When it comes to seeing such patterns outdoors 'in the field', the influence of abiotic factors is often most clearly seen where mountains and crags lead to large differences in the physical environment in a small area. Climate also tends to change as you move towards the poles – so that, for example, the wildlife of the south of England is rather different from that of northern Scotland. However, as an increase in altitude of 100 m is likely to reduce the temperature by over 0.5 °C, while a northward shift of one degree latitude (approximately 111 km) gives around 0.75 °C of cooling (Nagy & Grabherr 2009), it is much easier for a naturalist on the ground to see changes with altitude than with latitude (Fig. 17).

FIG 17. Tower Ridge on the north face of Ben Nevis. The difference in average annual temperature between the summit of Ben Nevis and the nearby town of Fort William at sea level is approximately 8.5 °C. This temperature difference is based on 13 years of data from the first half of the twentieth century (Pearsall 1950); using the 'rule of thumb' given in the text for temperature decline with altitude – of approximately 0.5 °C cooling per 100 m gain in altitude – gives a value of 6.7 °C. As is so often the case, the particulars of any one example depart somewhat from the predictions of a general theoretical model. In this case the rule of thumb (technically, the dry adiabatic lapse rate) makes particular simplifying assumptions about the moisture content of the atmosphere.

In the mountains such changes can be seen in a day's walk, while for latitude they require a day's drive. Differences between sites in the sun or full shade can be even more dramatic over very short distances. One implication of the nature of mountain climates is that organisms living in mountains may be somewhat buffered against the effects of climate change, having to move less far to find suitable conditions in a changing climate than if they were living in the flatlands (Loarie *et al.* 2009) – although there is obviously a danger that if the climate changes enough they could be left trapped on warming summits with nowhere cooler to move to.

A good example of the importance of topography is Winnats Pass in the Peak District. This is a narrow steep-sided limestone gorge just west of Castleton with a small road at the bottom, often congested with a mix of cars and sheep. As the middle of the pass runs almost exactly east–west, the steep slopes of the sides face due north and south. The north-facing slope here doesn't receive any direct sun before midday even in summer, and gets no direct sun at all between October and February. This gives considerable differences in microclimate between the two sides of the pass, and because of this it was used for a series of studies on the effects of the environment on vegetation and snails by scientists from the Unit of Comparative Plant Ecology at the University of Sheffield during the 1960s and early 1970s (Fig. 18). Much of this work was led by a then youngish scientist called Phil Grime, who went on to be one of the most influential British plant ecologists of his generation (Fig. 19). Grime and colleagues showed that the two slopes of the pass had different common larger snail species, with the banded *Cepaea nemoralis* being restricted to the grassy south-facing side, and *Arianta arbustorum* being found only on the north-facing side, which had areas of taller herbs such as Nettle *Urtica dioica* and Dog's Mercury *Mercurialis perennis* growing on it (Grime & Blythe 1969).

Part of the explanation for this is the difference in microclimate between the warmer south-facing slope and the cold north-facing one. For example, on a summer day in 1967, when Grime was doing this work, there was a difference of slightly over 4 °C in the maximum air temperatures recorded on the two slopes – that's roughly the difference one would expect to measure after travelling over 500 km north if you were comparing sites with similar altitudes and microclimate effects. In their 1969 research paper, published in the *Journal of Ecology*, Grime and Blythe pointed out that *C. nemoralis* appeared to be better adapted for resisting desiccation than *A. arbustorum*, as it much more readily formed a sheet of dried mucus (an epiphragm) across the opening of the shell, which helped prevent drying out. So there appeared to be obvious differences in the way the two species responded to the climate-related aspects of the abiotic

FIG 18. Winnats Pass, showing the shaded north-facing slope and the sunny south-facing slope. Work by David Pope as part of his PhD research in the early 1970s quantified the lack of light that the north-facing slope received – with no direct sunlight at all between October and February (Pope & Lloyd 1975). This photograph was taken at 11.30 am on 7 March 2017 and shows the sun only just appearing over the top of the south (north-facing) side of the pass. Almost 20 years after his PhD work on plant growth and topography in the Peak District, Pope was one of the supervisors of my own PhD work on light and urban trees.

environment, helping to explain their distribution. In addition, the *C. nemoralis* snails were mainly active by night, when it was cooler, reducing their exposure to heat and/or water stress.

Such microclimate effects, where factors such as the orientation of a slope or the presence of shading plants can locally modify the climate experienced by an organism, can be very important, allowing organisms to survive in favoured locations north (or south) of their normal geographic range. For example, while in Britain we think of Small-leaved Lime *Tilia cordata* as a mainly lowland tree species, at the southern limit of its range in the south of France it is found up in

FIG 19. J. Philip Grime photographed in 2012, some 45 years after his early work on snails and vegetation in Winnats Pass.

the hills, and then mainly on steep north-facing slopes or in the shade of ravines where it is less water-stressed (Pigott & Pigott 1993).

Grime's work on these snails is half a century old, raising the question of what their distribution in the pass looks like today. So on a late winter's day in early March 2017 I returned to Winnats to see for myself – curious, as the time span between Grime's work and today is effectively my life span. Had the snail distribution changed since I was about to start primary school? Patches of snow were still lying on the higher hills of the Peak District, but in the pass hints of spring were appearing, with Skylarks *Alauda arvensis* singing over the nearby fields and Jackdaws *Coloeus monedula* calling around potential nest sites on the crags. It was rather early in the year for a snail hunt, but the south-facing slopes were covered in empty *Cepaea* shells. Crossing the road onto the north-facing slope, I came across a few more shells, but not as many as on the sunny south-facing slope. Even when I hiked up the north-facing slope until I was some distance – as the snail crawls – from the sunny slopes, a few *Cepaea* could still be found (Figs. 20 & 21). So although the south-facing slope still seems best for *Cepaea*, unlike in the 1960s you can now find some on the cooler north-facing slopes too.

FIG 20. Empty *Cepaea nemoralis* shells collected during quick searches of approximately five minutes each on both the north and south sides of the road in Winnats Pass in March 2017. Those to the left of the scale bar came from the sunny south-facing slope, and those to the right from the shaded north-facing one.

FIG 21. View looking due north from part way up the north-facing side of Winnats Pass in early March. Even at this height I found the occasional *Cepaea* shell on the north-facing slope, where they had been completely absent in the 1960s.

Perhaps this change is a signal of a warming climate – certainly many British species have been moving north (and by implication also into once cooler microclimates) in the last 50 years. A short snail hunt cannot convincingly demonstrate the effects of climate change (at any site all sorts of other local factors may be responsible for the changes you see), but we have rather more convincing data from a range of well-studied species. This is particularly obvious for creatures such as butterflies, where Britain's great resource of serious amateur naturalists (now often rebranded 'citizen scientists') have amassed huge amounts of data. For example, the Speckled Wood *Pararge aegeria*, Brimstone *Gonepteryx rhamni* and Gatekeeper *Pyronia tithonus* have all expanded their British ranges northwards during the latter part of the twentieth century, at a time when the long-running data set on the temperature in central England suggests an increase in 1.5 °C over the last 25 years of the century (Asher *et al.* 2001). This matches my own experience too. I was a schoolboy naturalist around Manchester in the late 1960s and 1970s, and at that time you didn't see Speckled Woods, but they are now common in the area. Snails may be the epitome of slow-moving animals, but they may be responding in a similar, if slower, way to the butterflies.

THE BIOLOGICAL ENVIRONMENT

In Winnats Pass, the microclimates of the two slopes are not just affecting the snail species; they also have effects on the plants, and indeed on all the other organisms living there. During Grime's study the south-facing slope was mainly covered by vegetation consisting of low-growing grasses, with taller herbs – such as Nettles and Dog's Mercury – found on the north-facing slope where water loss was less of an issue. On my March snail hunt it was obvious that the south-facing slope was still largely grassy, while the north-facing one was damper, mossier and had more herb species amongst its vegetation. The *Arianta arbustorum* fed on these taller herbs, especially Nettles, while on the south-facing slope *Cepaea nemoralis* was feeding mainly on dead grass leaves. Feeding experiments under laboratory conditions carried out by the Sheffield group, using snails collected from the south-facing slope, showed that most fresh grasses were unpalatable to *C. nemoralis*, which was presumably why they appeared to be mainly eating dead leaves (Grime *et al.* 1968), and data from snails in the wild showed similar results. If you are wondering how the feeding behaviour of the snails was studied, it was by examining lots of faeces from wild-caught snails under the microscope. The faeces from *A. arbustorum* were green and contained the remains of Nettle stinging hairs, while the grassy remains in the *C. nemoralis* faeces were lacking green cell contents. So it appears that as far as the snails are concerned it's both the physical (abiotic) environment and the presence of food plants (which was in part determined by microclimate effects on plant growth) that are important.

In addition to the abiotic environment, ecology textbooks traditionally also discuss the concept of the biotic environment – those aspects of an organism's environment made up of other organisms, be they members of other species (such as the snail's food plants) or members of the same species (e.g. competition between individual snails for food plants or safe resting places on the underside of rocks). This distinction between abiotic and biotic aspects of the environment has been a well-used classification for many years. In his classic 930-page account of British vegetation published in 1939, Arthur Tansley devoted several introductory chapters to aspects of the abiotic environment (climate, geology and soils) before introducing the 'biotic factor', particularly the effects of grazing by animals.

Back in Cwm Idwal is a classic example of an experimental study of grazing and upland vegetation. In 1957 two exclosures were established in the Cwm as part of a study on the impacts of sheep grazing on the mountain vegetation (altogether nine exclosures were built across northern Snowdonia). An exclosure is simply a backwards enclosure: while an enclosure is designed to keep things in, an exclosure is designed to keep things out. It's the standard design for experiments to study the effects of herbivores on vegetation – exclude the herbivores and see what happens to the plants. In this case the fences were designed to exclude sheep – and probably also feral goats (Fig. 22), although these are much more agile than sheep – but herbivores of the size of voles and smaller still had access. Both exclosures in Cwm Idwal are still being maintained over 60 years since they were started. The one at the northern end of the Cwm is on rather acidic peaty podzol soil, while the other sits below the cliffs at the southern end of Llyn Idwal on a less acidic brown earth soil (there is also a much larger exclosure between these two, which was created as a management intervention to try and revegetate an eroding area, rather than for experimental reasons). Initially there was a complex experimental design with fences moved during the course of the year, so parts of the exclosure never had sheep grazing, while other parts were only grazed at particular times of the year. This was a rather time-consuming experiment to manage, as fences had to be repeatedly moved throughout the year, so since 1982 sheep have been excluded from the whole exclosure all year round (Hill *et al.* 1992). If designing such experiments today, one would ideally have many more replicates – the greater the number of exclosures the more confidence you can have in the results. However, the Cwm Idwal exclosures have the advantage that they have been running for a long time. This is a common problem with ecology field experiments; those that were set up many decades ago have the advantage of running for many years, but are often poorly replicated by modern standards of experimental design (other examples of this are the Park Grass Experiment discussed in Chapter 11, and the Godwin Plots described in Chapter 9).

The relatively high twentieth-century levels of sheep grazing helped create a grass-dominated environment in these North Walian mountains. However, there were clues that reducing grazing might lead to a different vegetation, as on top of some larger boulders at the back of the Cwm could be found heather-dominated vegetation where sheep had been unable to reach the plants. The same can be seen in the northern experimental exclosure, which now has many Heather *Calluna vulgaris* and Bilberry *Vaccinium myrtillus* bushes, although the difference between inside and outside the exclosure is now not as dramatic as it used to be (Fig. 23). Twenty years ago there was no obvious sign of Heather outside the

FIG 22. Feral goat in Cwm Idwal. Since the reduction in sheep numbers this is the commonest large herbivore in the Cwm – it was photographed in 2017 and was one of a herd of 22 individuals. Accepted practice is that domestic animals (and escaped feral stock) are given a different scientific name than the species which was originally domesticated. So the sheep in Cwm Idwal are *Ovis aries* and the feral goats *Capra hircus*.

FIG 23. The northern of the two exclosures in Cwm Idwal. The top photograph was taken in 2002, while the bottom one was taken in 2019. Note the much shorter grass outside the exclosure in the earlier photograph, with the Heather *Calluna vulgaris* bushes restricted to inside the exclosure. Sheep were excluded from the Cwm Idwal National Nature Reserve in 1998, but in 2002 any Heather outside the exclosure had not yet recovered from grazing and was still extremely small, so that it could only be found by a careful search amongst the grasses. Since then, Heather has started to grow taller outside the exclosure too. A comparable photograph from 2008 is reproduced in Sherratt & Wilkinson (2009), and one from 2015 can be found in Wilkinson & Sherratt (2016).

exclosure fence until you got down on hands and knees and started searching between the grass plants, where you could find very small Heathers hiding from the grazing sheep. These plants are now starting to grow larger following attempts to exclude all sheep from the Cwm in the late 1990s. The reduced sheep numbers have had dramatic effects – come here in July, and the area below the exclosure can now be yellow with flowering Bog Asphodel *Narthecium ossifragum*, making it look at first sight rather like a buttercup meadow. The long-term reduction in grazing caused by the fences hasn't substantially altered the species list for the exclosures but has allowed different species to become common (Hill *et al.* 1992, McGovern *et al.* 2014).

Clearly, greatly reducing sheep grazing in these exclosures has caused changes in the structure of the vegetation which most naturalists would see as an improvement, including the arrival of a small number of Rowan *Sorbus aucuparia* trees. These are presumably dispersed by birds which have fed on their berries, a mode of dispersal suggested by the observation that in several of these Snowdonian exclosure sites small Rowan trees are now growing next to fence posts – the obvious assumption being that these posts have been used by birds as perches, from which they have excreted or regurgitated Rowan seeds.

The fact that there are two exclosures in the Cwm, only around 1 km apart, also nicely illustrates how even in one nature reserve the same management change (in this case, removing sheep grazing) can lead to rather different results. The southern exclosure (Fig. 24) contains much less Heather. It's located on a 'richer' brown earth soil, and on occasions you can see the difference in soil with the help of Moles *Talpa europaea* – it's not unusual to see Mole hills around this southern exclosure, but you don't see them up at the northern one. In addition, there are small mountain streams running either side of the southern exclosure which can potentially flood the experiment following heavy rain. Mark Hill and colleagues (1992) described a 'small part' of this exclosure being covered with gravel following such a cloudburst in 1983. These two exclosures provide a nice example of how local-scale variation in soil type and disturbance events can cause a different outcome when sheep grazing is removed. A similar story of local variation in the environment having a noticeable effect on the vegetation can be seen in the two exclosures, from the same 1950s study, at the northern end of Llyn Llydaw on Snowdon, only a relatively short walk from the congested and expensive car park at Pen-y-Pass. As at Cwm Idwal, one exclosure has much more Heather than the other.

The effects of altering sheep grazing on the vegetation in Cwm Idwal are easy to see in the field and provide a nice illustration of the biotic environment as engineered by the grazing regime. Yet, much of biodiversity is invisible, being

FIG 24. The southern exclosure in Cwm Idwal, viewed from above, with streams flowing either side. Note that the vegetation here is very different from that found in the northern exclosure (Fig. 23). Photograph taken in 2017.

too small to see with the naked eye. Microorganisms are crucial to the working of ecosystems and are a key part of the biotic environment for the larger organisms more familiar to a field naturalist. As many gardeners and farmers are aware, the nature of the microbes in the soil (be they beneficial or not) is often crucial to plant growth. So what has been the effect of removing sheep and the subsequent changes in vegetation on the soil microbes? A study in the 1990s found that there had been a significant reduction in total microbes in the exclosure soils. The suggestion was that this may have been due both to the removal of nutrients from sheep urine and faeces, and potentially to changes in chemicals released from plant roots as the vegetation changed. These reductions in total microbial amounts, along with decreases in active soil fungi and nematode worms, were more pronounced on the brown earth soils (Bardgett *et al.* 1997).

A more recent study looked at the effects on nitrogen levels in the soils in these exclosures (McGovern *et al.* 2014). The removal of sheep grazing has caused the soils to become more nitrogen-rich, despite the vegetation changing to a more heathland type which would usually be associated with nitrogen-poor soils. So although removing grazing has apparently beneficial effects on the vegetation from a nature conservation perspective, this is not seen in the soils on a timescale of over half a century. Indeed, the removal of sheep has increased the amount of nitrogen compounds in the soil available to be washed into Llyn Idwal. So with sheep removal now being carried out over the whole Cwm it's possible that this will lead to increased nutrients entering the Llyn, with potentially negative consequences for some of the organisms that live in the lake. The authors of this 2014 study speculated that perhaps a more phased reduction in sheep numbers might give more chance for the soil ecology to change in a way that matches the changes in the above-ground vegetation.

THE BLURRING OF ABIOTIC AND BIOTIC ENVIRONMENTS

The complexities in changes to soil microbes nicely illustrate a problem with the simple distinction between abiotic and biotic environments that has been used in many textbooks, and also in this chapter so far. Changes in sheep numbers alter the nutrients entering the soil (chemistry – so traditionally considered abiotic, although an animal is involved in this case), and the behaviour of the microbes (biotic) has the potential to alter the lake chemistry too. The apparently obvious distinction between the biological aspects of the environment and its physics and chemistry is, at best, less distinct than it looks at first glance. These aspects of the chemistry of the environment are related to the life that lives there.

The broad flat-bottomed glacial valley below Cwm Idwal may also hint at a similar story. It's called Nant Ffrancon (Fig. 25), which is often translated as 'Beaver's hollow', although other non-zoological translations are possible, and it's also plausible that it may be named after a confused reference to Otter *Lutra lutra* rather than Beaver *Castor fiber* (Aybes & Yalden 1995). Using place names as data in natural history is often difficult! However, Beavers were formerly widespread in Britain and are starting to slowly reappear due to reintroduction schemes – both official and clandestine. Because of their well-known ability to fell trees and use them to dam watercourses to create deeper water for their own use, Beavers are classic examples of an organism altering something that we would normally consider part of the abiotic environment, namely water depth (Fig. 26). The current near absence of trees in Cwm Idwal and Nant Ffrancon is down to animals too, in this case ourselves and our domestic animals. A summary of woodland cover in a range of Welsh sites around 8,000 years ago – based on fossil pollen evidence from sediment cores – suggested that Cwm Idwal and Nant Ffrancon were two of the more heavily forested locations in the study (Fyfe 2007).

FIG 25. The Nant Ffrancon valley, with Cwm Idwal in the shadows upper left, below the snowy summits. The name of the valley plausibly refers to Beavers *Castor fiber*. Certainly Beavers were widespread across Britain in the past. Nant Ffrancon is a classic wide flat-bottomed glacial valley which contained a lake at the end of the last glaciation, and analysis of pollen grains preserved in the sediments from the former lake shows that 8,000 years ago the surrounding hillsides were mainly forested (Fyfe 2007).

Fossil pollen also shows that the presence of the arctic–alpine plants in Cwm Idwal goes back to the retreat of the last ice sheets, and that they hung on through the more forested condition – presumably on crags too steep to support trees (Tipping 1993).

This chapter started with the thought that the more extreme environment of the British uplands, along with the lower number of species found there, may help clarify ideas on the role of the environment in ecology which may be harder to see in more species-rich lowland habitats. The effects of the abiotic environment are clearly seen, and obvious to any naturalist with the energy and fitness to walk up a mountain. This is nicely illustrated by *A Flora of Cumbria* (Halliday 1997), which gives the highest altitudinal record in Cumbria for each plant species listed. For example, some species are restricted to relatively low altitudes, such as Dog's Mercury (maximum altitude 460 m) and Bog Rosemary *Andromeda polifolia* (490 m). Plants making it to high on the fells include Lesser

FIG 26. Pond caused by flooding from a Beaver dam. The activities of these animals alter the physical environment for many other species – note the trees that have died when their roots were flooded. These dead trees in turn provide habitat for many invertebrates and fungi. Because of their ability to alter the environment for many other species, Beavers are often referred to as ecosystem engineers (discussed in more detail in Chapter 10). This photograph was taken in Canada and shows a pond constructed by the North American Beaver *Castor canadensis* (a different species from the Eurasian Beaver – although in the same genus). It is sometimes claimed that the American species constructs more dams than the Eurasian one, although there is no consensus on this in the literature on the natural history of these two species (Busher 2016).

Clubmoss *Selaginella selaginodes* (885 m), Purple Saxifrage (885 m) and Starry Saxifrage (915 m). For these two saxifrage species (Figs. 14 & 15) the Flora also gives the lowest altitude for Cumbria, as they are not found down to sea level. For Purple Saxifrage this is 245 m, while for Starry Saxifrage it's 130 m. Presumably at lower altitude these arctic–alpines tend to be outcompeted by other plants less able to tolerate the conditions high on the mountains – and indeed this speculation is supported by recent experiments carried out in the Swiss Alps. As part of an attempt to understand the possible effects of future climate change, several alpine plant species were transplanted to sites lower on the mountainside, where their growth was 'strongly reduced' by competition from the surrounding lower-altitude plant species (Alexander *et al.* 2015).

These competitors are an aspect of the biotic environment. The Snowdonia exclosures provide another nice experimental example of the importance of the biotic environment – in this case sheep grazing, but in much of the Scottish Highlands grazing by high numbers of Red Deer *Cervus elaphus* can be important too (Moore & Crawley 2015). However, as the examples of the Welsh exclosures and the Beaver show, this split into abiotic and biotic when thinking about an organism's environment needs using with care, as not only do the physics and chemistry of the environment affect what lives at a location, but life can affect the physics and chemistry of the environment too. On shorter timescales, separating the environment into physics/chemistry or biology can sometimes be a useful way of thinking about things, but on longer timescales the distinction becomes ever more unhelpful – after all, almost all the oxygen in the atmosphere has come from oxygenic photosynthesis (that is, oxygen-producing photosynthesis – other forms of photosynthesis are used by some microorganisms). However, if all photosynthesis were to stop today it would take around 4 million years for the atmosphere to run out of oxygen (Lenton 2016). On the short timescales of natural history fieldwork, atmospheric oxygen can usefully be considered as a stable part of the abiotic environment, even though on a geological timescale this is clearly not correct.

CONCLUDING REMARKS

The British uplands in general, and Cwm Idwal in particular, illustrate a range of basic ideas about the nature of an organism's environment. In introducing these ideas to students it's been common to split the environment into two types, the abiotic (physics and chemistry) and biotic (other organisms). The often harsh climatic conditions in mountains make the abiotic aspects of the environment particularly obvious to the naturalist. Not only are the general climatic conditions important, but also mountains with their steep slopes and rock faces clearly illustrate the importance of microclimate in species distribution, as seen in the examples from Winnats Pass in the Peak District. For example, although locations may be sometimes only metres apart, north- and south-facing slopes may experience very different local climates. One long-standing theoretical approach to thinking about how such environmental factors affect an organism's distribution and population is the 'law of toleration', which describes the effects of such factors on a species' success and survival. It suggests that for a factor such as air temperature, or the concentration of nitrogen in soils, there is for any organism an optimum value that can support good numbers of that species, along with maximum and minimum values beyond which that species cannot survive. Such ideas can form the starting point for mathematical treatments of the way an organism's distribution may be altered by things such as a changing climate.

In the British uplands past grazing pressure from sheep has also made it easy to visualise the biotic environment in action, as shown by the Snowdonia exclosure experiments. This also illustrates the way humans have affected the upland vegetation through agricultural practice, such as sheep stocking density. However, this binary abiotic/biotic classification is too simple, and indeed sometimes very misleading. As the examples of the microbiology of the Cwm Idwal exclosures and the past (and future?) effects of Beavers in the valleys show, this split between abiotic and biotic is rather artificial, albeit useful on occasions. It has long been known that abiotic factors such as climate affect organisms, but many aspects of the physics and chemistry of the environment are in part determined by the actions of life itself. Organisms which affect the abiotic environment by their physical presence (such as the shade cast by a tree) or their actions (such as a Beaver building a dam) are sometimes referred to as 'ecosystem engineers' because of their role in constructing the environment. Indeed in some cases, such as photosynthesis and oxygen in the atmosphere, the effects of life are crucial. The microbes in the soils of Cwm Idwal also illustrate the important fact that much of biodiversity is too small to see with the naked eye, but these microbes play fundamental roles in many ecological processes such as nutrient cycling. The importance of microbes in ecology is an idea that will be returned to repeatedly in various chapters of this book.

CHAPTER 3

Wytham: Questions about Life in a Deciduous Woodland

Autumn in Wytham Woods, just west of Oxford. Many of the leaves have already fallen, allowing the low-angled sun to penetrate to the woodland floor. The older Beech trees stand out dramatically, still having most of their leaves – yellow with carotenoid pigments in the autumnal sunshine (Fig. 27). Although it is a very long way from anything that could be called wilderness, or even wild, there is a measure of naturalness about the wood. With the wind blowing through the trees, even the traffic on the Oxford ring-road is largely inaudible, an absence of the racket of human life that is depressingly rare in much of southern England.

To an ecologist, Wytham is a special place. As you walk round the wood the reason for this starts to become apparent. Scattered throughout the trees are the paraphernalia of field ecology: marker canes and here and there a thicker post indicate the position of experiments and survey points, numbered bird nest boxes abound, as does a range of insect traps, and tree trunks are encircled by bands to record their increase in girth as they grow (Fig. 28). Away from the public paths are signs of more substantial scientific infrastructure – a flux tower for atmospheric measurements, and a scaffolding walkway to allow access to life in the canopy (Fig. 29). As Oxford biologist John Krebs (2010) has written of Wytham, 'It is almost certainly unmatched anywhere in the world as a place of sustained, intensive ecological research extending over nearly three quarters of a century.' The explanation for this long history of use in research is the donation of the woods to the University of Oxford in 1942. At first it was mainly used by the Forestry Department, but from the 1960s the emphasis increasingly became research in ecology and animal behaviour (Savill *et al.* 2010). Krebs' own

FIG 27. Autumn Beeches *Fagus sylvatica* in Wytham Woods.

FIG 28. A dendrometer band in Wytham Woods, used to measure the expansion of trunk diameter over time. These bands are made of plastic tape, fixed around each trunk, and fastened with a spring. As the tree grows, the end of the band moves away from the original starting point and the distance can be recorded to give a measure of increase in diameter. Such studies show that weather can have an important role in determining rate of growth: in Wytham, for example, low temperatures in 2010 were associated with a 40 per cent reduction in growth rate compared to 2011 and 2012 (Butt *et al.* 2014).

attachment to Wytham dates back to his doctoral research in the woods during the 1960s – and indeed, on entering the House of Lords in 2007 he took the title Lord Krebs of Wytham.

The importance of Wytham to ecological research explains why several of my own visits to the woods have been on field trips associated with academic conferences. Ecologists from other parts of the UK and the rest of the world are often keen to see this classic site for themselves. The peak of the woods' scientific, and textbook, fame comes from the mid-twentieth century when several 'great men' of ecology and animal behaviour (and they did tend to be men in those days) were based at Oxford and used the woods as an important research site. In particular Charles Elton, one of the founding figures of animal ecology, led a group that studied the small mammals in the woods (Fig. 30), while David Lack and his associates studied the birds, especially the Great Tit *Parus major* and Blue Tit (see Chapter 10). Starting with work by Hans Kruuk in the 1970s, the woods'

FIG 29. Canopy walkway constructed in Wytham Woods in 1993 to allow access to the tree canopy (photographed in April 1999). This is where most of the trees' photosynthesis takes place, and it is also rich in invertebrates such as Winter Moths *Operophtera brumata*, a key food for many insectivorous birds. Because of the difficulty in access, woodland canopies are understudied by ecologists.

Badgers *Meles meles* have also formed the basis of an important long-term study (e.g. Macdonald *et al.* 2004, Buesching *et al.* 2010).

Interestingly, another key Oxford figure of the mid-twentieth century didn't take to Wytham as a research site. The animal behaviourist – and later Nobel Prize winner – Niko Tinbergen spent quite a bit of time walking and taking photographs in the woods during his first few years in Oxford, but 'he was not drawn to study any of its denizens' (Kruuk 2003). Perhaps as a landscape it was too different from the coastline of the Netherlands where he grew up and carried out his early studies. Later it was the sand dunes around Ravenglass in Cumbria that were to become his key British field site. For a scientist with a natural history

FIG 30. The Elton memorial in Wytham Woods. Charles Elton was a key figure in the history of animal ecology. He was based at the University of Oxford and made extensive use of Wytham from 1942 (when it was left to the university by Colonel Raymond ffennell). Elton's field diaries from 1942 until 1965 recorded around 10–15 visits per year and now form an important record of changes in the woodland in the mid-twentieth century (Kirby 2016).

background there can be more to picking a field site than the mere practicalities of a particular research project. Some sites work emotionally and aesthetically as well as scientifically.

In woodland, several key ecological processes are particularly easy to visualise – and this is true for any type of deciduous woodland, from lowland woodlands, like Wytham, to the moss-rich upland oak woods in the mountains (Fig. 31). The idea that photosynthesis forms the basis of most food chains is familiar from early school science lessons. Photosynthetic green plants are particularly obvious in woodland, and the size of some of the larger trees dramatises important ideas about how photosynthesis works. In addition,

FIG 31. Deciduous woodland in the British mountains, in this case in the Cadair Idris National Nature Reserve, Snowdonia. Note the extensive mosses growing on the woodland floor in this rather wet part of the country – indeed, it is often described as temperate rainforest.

the annual loss of leaves in the autumn raises a crucial question – albeit one often overlooked because of its familiarity – *where do all those leaves go?* Why are deciduous woodlands not clogged with great drifts of dead leaves, as one year's leaf fall stacks up on those from previous years, decades and centuries? The answer of course is decomposition, and there is nowhere better to appreciate its importance than in deciduous woodland. This chapter develops these ideas, mainly illustrated by examples from deciduous woodlands, supplemented with a number of marine examples, as these are universal ecological ideas applicable to almost all habitats. Because Wytham has been so intensively studied it is often the case that this one wood just outside Oxford can provide the data needed to make these ideas more quantitative – to put numbers to the processes that can be seen in any deciduous woodland, or indeed any other habitat. So studies carried out in Wytham feature repeatedly throughout this chapter, and in Chapter 10 too.

PHOTOSYNTHESIS – EATING THE SUN

Imagine your favourite really large tree – perhaps one of the ancient oaks of Sherwood Forest in the Midlands, be it the tourist magnet of the Major Oak or another of the venerable examples scattered around the National Nature Reserve (Fig. 32). The history of past management at Wytham means that it can't match Sherwood for large numbers of physically impressive really old trees, for in much of Britain really large old trees are most likely to be found at sites that were more open in the past. At Wytham many of the older larger trees are old field trees now incorporated into the woodland. Historically much of Sherwood Forest was heathland with trees, hence the spectacular collection of ancient oaks (Rackham 2003).

Any large tree prompts a number of questions – what is it made of, and where has all the material that made that timber come from? Given that most people know that trees have roots, and draw some sort of sustenance from the soil, the obvious (if wrong!) answer is that they are in some way eating soil. However, a moment's reflection is likely to reveal a major problem with such a theory. If a large tree were mainly made from soil, surely there should be big holes in the ground under a really impressive tree where all the soil used to be that has now been converted to trunk and branches? The real answer is far more remarkable, as most of the solid bulk of a big tree (or indeed any other green plant) is made of air. That is worth repeating for emphasis – trees are made of air. The basic idea is introduced to most people in school science, although sadly often as a set of facts to learn by rote, without the extraordinarily counterintuitive 'made-of-air' aspect being emphasised. It's all there to see in the simple summary equation of photosynthesis (to be technically exact, 'oxygenic photosynthesis', as will be described later). A key raw material is carbon dioxide in the air, with the main product being sugar (the simple carbohydrate glucose):

$$\text{carbon dioxide} + \text{water} \rightarrow \text{carbohydrate} + \text{oxygen} + \text{water}$$

Light energy from the sun is used to power parts of this reaction, so in the evocative phrase of the science writer Oliver Morton (2007), the process of photosynthesis amounts to 'eating the sun'. It should be pointed out that the simple chemistry equation above is just a summary of what happens overall, and photosynthesis is really composed of a complex series of chemical reactions. Although water appears on both sides of this summary equation it is not the same water molecule. The water on the left-hand side of the equation is broken down so that its constituent parts can be used, and 'new' water is later formed (as seen on the right-hand side of the equation).

FIG 32. The large old oak trees of Sherwood Forest National Nature Reserve, Nottinghamshire. Top: the Major Oak is supported by extensive scaffolding and treated more as a historic monument than an actual living (and dying) plant. A Pedunculate Oak *Quercus robur* of this age would be expected to be falling apart and its dead wood providing habitats for a wide range of species. Bottom: Sherwood Forest has a large number of other old oaks, around 500 years old (Rackham 1990). Not quite as old or famous as the Major Oak, these are being allowed to fall apart naturally (within the constraints of visitor safety).

If the impressive solidity of carbon dioxide turned to solid plant is best seen in a woodland, another key aspect is better seen in a sunny pond or rock pool. In bright sunlight it is sometimes possible to see bubbles of oxygen rising from aquatic plants. The example illustrated (Fig. 33) was in a rock pool on Staffa in the Inner Hebrides; peering into the brightly lit water, I could see, on a small scale, an illustration of the origin of most of our atmospheric oxygen. The bubbles rising from the algae are oxygen created as a by-product of photosynthesis.

The realisation that most of the oxygen in our atmosphere comes from photosynthesis is relatively recent. The idea was developed during the mid-twentieth century; so some of the oldest readers of this book will have grown up at a time when we thought that most of the oxygen came from non-biological sources. Water is a molecule that comprises two atoms of hydrogen and one of oxygen (H_2O), and it used to be thought that much atmospheric oxygen came from water molecules in the atmosphere being smashed asunder by energetic ultraviolet light. One of the reasons this can't work is that as you make more atmospheric oxygen (in the form of the molecule O_2) you also set

FIG 33. A green alga of the genus *Ulva* (in older books it is placed in the genus *Enteromorpha*) photosynthesising in a brightly lit rock pool on the Hebridean island of Staffa. Note the bubbles of oxygen rising from the alga.

up the conditions for the formation of ozone (O_3), and, as is well known, the atmospheric ozone layer reduces the amount of ultraviolet light reaching the lower atmosphere. So increases in atmospheric oxygen would create conditions that reduce the amount of new oxygen that can be created by non-biological processes – hence the importance of photosynthesis for making most of our oxygen (Wilkinson 2006). Without this oxygen, almost all the organisms described and illustrated in this book could not survive, and the same goes for the author and his readers too! So photosynthesis is fundamentally important to most life on Earth – both as a source of oxygen and as a source of food.

The classic demonstration of the importance for life of what we now call oxygen was by Joseph Priestley in the 1770s. He showed that a mouse in a sealed jar would die, but that if he added sprigs of mint the mouse was fine. We now know that as the mint photosynthesised it was giving off oxygen and taking in carbon dioxide. The resources of television documentary making have allowed this experiment to be repeated on a larger scale. Priestley's jar was re-created as a large (2.0 × 2.5 × 6.0 m) airtight transparent box at the Eden Project in Cornwall which, when stocked with a wide range of plants, was able to support a human for a couple of days. Indeed, to make the demonstration more dramatic the system was started at lower than normal oxygen levels, and despite the use of oxygen for breathing by its human inhabitant the oxygen levels actually rose in the two days of filming during which the programme's presenter was incarcerated in the box (Martin *et al.* 2012).

Beyond supplying oxygen, photosynthesis is crucial in ecology for other reasons. For example, another bit of basic ecology most of us meet at school is the concept of the food chain, and often the base of a food chain is a photosynthetic organism making use of solar energy. For example, in a deciduous woodland such as Wytham a simple food chain could look something like this:

oak leaf → winter moth caterpillar → great tit

The arrows show the flow of energy along the food chain, as solar energy is converted into oak leaves that are then converted to caterpillars before these caterpillars are converted into birds. So much of the energy powering life around us comes either directly or indirectly from the sun via photosynthesis. It's worth pointing out, however, that while this familiar oxygenic (i.e. oxygen-producing) photosynthesis is crucial to much of life on Earth, not all photosynthesis is of this oxygenic type. Some groups of photosynthetic bacteria use other mechanisms to 'eat the sun' which don't produce oxygen as a waste product; examples include purple bacteria and various types of sulphur bacteria.

DECOMPOSITION

A deciduous woodland in autumn also spectacularly illustrates another key idea in ecology – decomposition. Year after year, leaves fall from the trees to the woodland floor, and yet there is no steady year-on-year increase in the depth of leaf litter covering the soil. You don't walk into a wood such as Wytham and wade waist-deep through vast piles of dead leaves (Figs. 34 & 35). This is

FIG 34. Deciduous woodlands change dramatically during the year with the loss of leaves in autumn. The two photos show Bunny Old Wood in Nottinghamshire in late September 2017, just before the start of autumnal leaf loss, and in February 2018, with no leaves.

FIG 35. Leaf litter in Tattershall Carrs Wood in Lincolnshire. This highlights the importance of decomposition – leaf litter doesn't build up year on year but is broken down by a range of organisms, thus making the energy and nutrients in the leaves available to other organisms.

something so familiar that we seldom think about it, but it is of tremendous importance. Clearly the reason why the forest floor isn't metres deep in leaf litter is that these leaves are broken down – decomposed – by a wide range of organisms.

The huge scale of what's going on can be hard to visualise when looking at a relatively natural system such as a woodland floor. But it can be easier to see in a more managed system where humans are clearing up the leaves to keep things 'tidy'. This was brought home to me when I sat for a while in a London square on my way home from a scientific meeting. The gardens in Russell Square, Bloomsbury, were designed in the eighteenth century by the famous landscape gardener Humphry Repton, and now contain many mature trees. As I sat in the gardens on a sunny autumn morning before going to catch a train home from Euston Station, a team of gardeners were busy collecting the fallen leaves and filling large bags with the results of their work (Fig. 36). Seeing the leaf fall of trees in a small London square collected together like this illustrated just how much material is broken down each year in a deciduous woodland. Indeed it is not just deciduous woodlands or urban parks where the extent of

FIG 36. Russell Square, London, in autumn. When collected together by gardeners, the quantity of leaves that fall from trees in autumn becomes very obvious.

the material available for decomposition can be readily appreciated – strand lines on beaches make the point almost as well (Figs. 37A & B), at least when not tidied up to keep the beach 'clean'. There is a lot of dead organic matter to be broken down in the world.

The organisms that break down leaf litter, and other organic remains, are effectively doing the reverse of photosynthesis. As the simple summary equation of oxygenic photosynthesis shows, carbon dioxide is removed from the air (or water, in aquatic systems) and turned into plant material. This plant material is sometimes then converted into animal if eaten by herbivores, which in turn can be eaten by predators – the classic idea of a simple food chain. That is, organisms higher up the food chain make use of energy from photosynthesis by eating plants or eating organisms that have fed on plants. Respiration – the way many organisms release energy from carbohydrates within cells (be they in plants

FIG 37A. While particularly apparent in deciduous woodland, decomposition is a general process occurring in all habitats. In this example decomposition is illustrated by a freshly dead Gannet *Morus bassanus*, with another partly decomposed Gannet shown in Fig 37B.

or other organisms) – is effectively the reverse of photosynthesis and can be summarised as:

$$\text{carbohydrate} + \text{oxygen} \rightarrow \text{carbon dioxide} + \text{water}$$

In this biochemical context respiration refers to the chemical processes within the cell by which an organism extracts energy from food – rather than the act of breathing, which is the commoner use of the word in everyday English. In respiration, therefore, oxygen is being used and carbon dioxide given out. Of course plants, and other photosynthetic organisms, respire too. In a plant, however, the effects of photosynthesis (e.g. taking in carbon dioxide) usually outweigh the effects of respiration, such as giving off carbon dioxide, so that a plant as it grows can be thought of as an oxygen producer. This changes when

FIG 37B. A decomposed Gannet on a Scottish strand line.

the plant dies, and this equation also tells us something interesting about photosynthesis as a source of oxygen in the atmosphere.

Decomposition is the use of dead organisms via respiration to provide energy for decomposers, and so it uses up oxygen (in much the same way, predation is the respiration of living organisms). For there to be some oxygen left over in the atmosphere, then, some plant material must survive long-term in a relatively undecomposed form. Without this, the oxygen produced by the plant during its lifetime is consumed as it is respired (either by a herbivore or during decomposition). An important environment where dead photosynthesisers can survive undecomposed is to be found in the muds on the ocean bed, and this is why photosynthetic marine plankton are particularly important as a source of atmospheric oxygen.

So the burial of organic matter – such as dead plankton or plant material on the seabed – provides a long-term source of oxygen to the atmosphere (Wilkinson 2006, Lenton 2016). Plants on land are less likely to be preserved long-term in a relatively undecomposed form, although it can happen – for example, peat is mainly poorly decomposed plant material that can survive for thousands of years. However, a disproportionate percentage of the oxygen in the atmosphere comes from aquatic photosynthesis.

Much of this decomposition on land is being carried out by microorganisms, out of sight to a naturalist in the field. However, one important group of decomposers can be more visible, especially in autumn – namely the fungi, although not all of these are involved in decomposition. For example, some are parasitic while others form mutually beneficial associations with many of the woodland plants (these will feature in other chapters.) However, many fungal species are key decomposers, especially of wood. The familiar mushrooms and toadstools that we see are really just spore-dispersal devices, and the main body of the fungus is out of sight in the leaf litter or rotting wood. This is the mycelium, an often extensive network of microscopic fungal threads called hyphae (Fig. 38). When clumped together in a different arrangement, the hyphae also comprise the fruiting bodies that are the familiar mushrooms and toadstools.

One reason that fungi are important in woodland decomposition is that some of them can break down wood, a biochemical trick which is beyond the abilities of most organisms (Figs. 39 & 40). In particular they can break down lignin, a large molecule (chemically a complex polymer) that provides mechanical support in woody plants. The effects of lignin on decomposition have been nicely illustrated by a global experiment looking at the decomposition rates of small bags of two types of tea (green tea and rooibos) at 570 sites around the globe on every continent except Antarctica. This global reach requires a large team of scientists, and the research paper detailing the results lists 310 authors (Djukic *et al.* 2018)! The project is also distinguished by having a great pun as its name – it uses tea to study decomposition, and is therefore known as the 'TeaComposition initiative'. Rooibos contains more lignin than green tea, and as might be expected it did tend to decompose more slowly. Indeed the litter type (rooibos or green

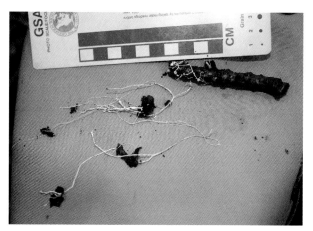

FIG 38. Fungal mycelium excavated from leaf litter in woodland at Meres Sands Wood nature reserve in northwest England. Note how it has colonised a dead twig, which the fungus is breaking down.

FIG 39. Fairy Inkcap *Coprinellus disseminatus* growing on a rotting stump at the Wildfowl & Wetlands Trust (WWT) Slimbridge reserve, Gloucestershire.

FIG 40. King Alfred's Cakes (also known as Cramp Balls) *Daldinia concentrica* growing on dead wood in Tattershall Carrs, Lincolnshire. This is a very common fungus, most often found on dead or dying Ash trees. There are several other similar-looking species of this genus living in Britain.

tea) was the most important influence on rate of decomposition, more so than climate – which only became really important in more extreme environments such as very dry conditions, where decomposition rates were greatly reduced. This neat experiment, using a simple and cheap methodology, has nicely illustrated the relative roles of litter quality and climate, which have been understood for some decades but have acquired new significance in the context of the potential effects of climate change on decomposition rates (Coûteaux *et al.* 1995). This is a potentially important issue, as any increase in decomposition rate with a warmer climate would release more carbon dioxide from the decomposition of organic matter in soils, which would have the potential to increase the rate of climate change.

Rotting wood is important not only in the context of nutrient cycling but also as a habitat for a range of species; as well as the fungi, it supports a range of beetles, including some of Britain's rarest. The key requirement for these beetles appears to be a long-term availability of dead wood in various stages of fungal decay (Jones 2018). Because dead wood has often been tidied up in the British countryside, such sites can be infrequent and there is now a conservation interest in providing more of this habitat, ranging from people leaving small piles of dead wood in the corners of their gardens to allowing the remains of large trees to decay naturally in woods and parklands (Fig. 41). Indeed, because of its long history of woodland management and associated tidying, Wytham is mainly lacking in specialist dead-wood invertebrates (Hambler *et al.* 2010).

What influences the rate at which wood is decomposed by fungi? The discussion in the previous chapter introduced the idea of the importance of microclimate, using examples from the British uplands. Studies at Wytham Woods show that microclimate matters here too. In an experiment on wood decomposition rates, fungi broke down small blocks of Beech wood placed on the woodland floor more slowly towards the edge of the wood, where the drying effects of greater exposure to sun and wind resulted in a lower moisture content. As so many British woods now exist as relatively small fragments, with much of the woodland affected by such microclimate edge effects, this has interesting implications for the rate of decomposition in these habitats. Martha Crockatt and Daniel Bebber (2015), who carried out the Wytham wood-block study, used their data to estimate that taking such edge effects into account would decrease the rate of decay by around a quarter in British woodland compared to the situation if there was no reduction in decay rate around the woodland edge.

Fungal decomposition of trees is also potentially important on a geological timescale. It is possible that the fungal ability to break down wood had a role to play in the industrial history of Britain, and much of the rest of the world too. Most coal deposits come from the late Carboniferous and early Permian (approximately 323–252 million years ago), as for some reason far more coal was deposited during this period than at other times. One possible explanation, worked out in detail by Jennifer Robinson some 30 years ago, implicated fungi. She suggested that the trees at the time – mainly tree-sized versions of clubmosses, ferns and horsetails – had higher lignin content than today and that the ability to break it down did not evolve in fungi until later (Robinson 1990). With much less breakdown of the dead tree material, far more would have made its way into wet swamp sediments, eventually to become coal.

FIG 41. Dead trees left as a conservation measure at the National Trust's Dunham Massey Park in Cheshire. The site was a medieval deer park, and the current landscape is largely seventeenth-century in origin. The long continuity of trees and dead wood makes this an important site for rare beetles, and other species, associated with rotting wood. Although a well-known site for its dead-wood insect assemblage, there is some evidence to suggest that it is not as good for these animals as the British woodlands of 5,000 or more years ago. Beetle remains extracted from modern pond sediments at Dunham do not contain as many remains of dead-wood insects as similar aquatic sediments from before the start of extensive woodland management in prehistoric times (Smith et al. 2010).

Recently these ideas have been challenged (Nelsen *et al.* 2016), and an explanation for the high rate of coal formation that is based on climate and the arrangement of the continents at the time has been suggested to be more likely. Certainly it looks increasingly unlikely that the large vegetation of the time was unusually rich in lignin, as had been assumed in the past, but a role for fungi still seems possible. Fungi leave few fossils, so much of our knowledge of their evolution comes from analyses of the genetics of modern fungi interpreted in the context of the rather limited fossil record. This can allow an estimate to be made of when the various types of fungi first evolved. Such an approach suggests that the lignin-degrading ability of some fungi arose around the right time to explain the reduction in coal after the mid-Permian; however, there are understandably large uncertainties attached to these analyses, so it is still possible that this ability of fungi evolved rather earlier – which would weaken the case for a lack of wood-degrading fungi at the time of peak coal formation (Hibbett *et al.* 2016). These genetic analyses also suggest that this lignin-degrading ability first evolved in the jelly fungi – probably the most well-known British example being Jelly Ear *Auricularia auricula-judae.*

The decomposition of leaf litter and other plant parts is effected not only by microbes and fungi but by many animals as well. If you mention the role of life in decomposition of leaf litter, earthworms will be among the first organisms to occur to many people. One such person was Charles Darwin (1881), who was particularly interested in the role of earthworms in soil formation and devoted his final book to the topic, championing these invertebrates by writing 'Worms have played a more important part in the history of the world than most persons would at first suppose.' However, it's not just worms that are potentially important in temperate forests such as Wytham Woods, as woodlice and millipedes are some of the most important larger invertebrates involved in decomposition in such a habitat.

One study at Wytham investigated the importance of these larger invertebrates by using litter bags of two different sizes (Riutta *et al.* 2012). Bags of leaf litter, as in the TeaComposition experiment already described, are a standard way of studying the breakdown of plant matter. The bags were filled with leaf litter of either Ash *Fraxinus excelsior* (which readily decomposes) or Pedunculate Oak *Quercus robur* (which is slower to decompose), and some had a mesh size that would exclude invertebrates of the size of woodlice and millipedes while others had a larger mesh size that would allow them to access the leaf litter. Sampling of the soils at the experimental sites showed that there were five woodlouse species and six millipede species, with the most abundant being the Common Striped Woodlouse *Philoscia muscorum.* The experiments ran for three months over the summer of 2009, with some litter bags placed near the edge of the wood and some much further into the wood. They also experimentally added water to some of their soils. For both the Ash and the Oak litter there was a higher rate of decomposition in the bags that allowed access to woodlouse-sized invertebrates, showing they have a role in the decomposition of leaf litter. In addition, as expected, Ash decomposed more rapidly than Oak and a higher moisture content increased the rate of decomposition. As with the fungal wood-block study at Wytham described earlier, decomposition was slower near the woodland edge, and this seemed to be because of lower moisture content. This pattern of faster litter decomposition in litter bags that allow in larger invertebrates is not exclusive to Wytham and is often observed. For example, it was found in a similar study carried out at Silwood Park near Ascot, Berkshire, and in this case several species of earthworms appeared to be important (Milcu & Manning 2011).

The Wytham data showed something else that is very informative. Although they established that the invertebrates are important, and also that increased moisture accelerated decomposition, there was no relationship between the invertebrates and moisture. The moisture appeared to mainly affect the microorganisms and fungi involved in decomposition, not the woodlice or millipedes. However, the presence of the invertebrates speeded up the decomposition rate by leading to the leaf litter being broken up into smaller fragments which were more readily broken down by the microbes. The earthworms studied by Darwin and also in the Silwood Park study have a similar effect. So the invertebrates are not *required* for leaf-litter decomposition, but they can speed up the rate at which it happens. This is probably typical of the role of animals in many important ecological processes. The analogy I find most useful comes from chemistry. Animals can be thought of as catalysts for many ecological processes, allowing the processes to happen more easily and more rapidly (Wilkinson 2006).

CARBON STORES

The complementary concepts of photosynthesis and decomposition come together when we start to wonder about the amount of carbon stored in a deciduous woodland. One consequence of the idea that plants are made of carbon dioxide is that trees and other plants are living stores of carbon. This is why there are such serious worries about forest destruction in the context of climate change, and why environmentalists often advocate tree planting as a method of reducing the size of climate changes associated with a human-driven increase in atmospheric carbon dioxide. Some of the carbon compounds created by photosynthesis are respired by the plant as an energy source – with an associated release of carbon dioxide to the environment. However, some of the carbon fixed by photosynthesis contributes to the body of the plant, and in a long-lived organism, such as a tree, may be taken out of circulation for many years.

So if you look at woodland such as Wytham you are, amongst many other things, looking at a carbon store – but much of the carbon is not actually stored where one might expect (Fig. 42). This is nicely illustrated by data from another well-studied deciduous woodland, Monks Wood National Nature Reserve in Cambridgeshire (Patenaude *et al.* 2003). During 2000 the numbers and sizes of trees, shrubs, dead wood and so forth were recorded and their carbon content estimated by using previously published mathematical equations relating the size of a plant to its carbon content (these equations vary between different tree species). In addition, a series of soil samples was taken to a depth of 50 cm and the amount of carbon in the soil was measured. As you might expect, there was a lot of carbon locked up in the larger trees – some 78 tonnes per hectare. Larger trees were defined as those with a trunk diameter at breast height (i.e. 1.3 m above the ground) of 18 cm or greater. 'Diameter at breast height' is a standard measure used in forestry – it is easy to measure. The other plant material (dead wood, foliage, leaf litter, shrubs, ground vegetation and roots) contained around 54 tonnes per hectare; this seems no great surprise, as one might guess that the larger trees would dominate carbon storage in a wood. However, they didn't! In Monks Wood there was around 335 tonnes per hectare of carbon in the soil (excluding larger roots). So most of the carbon stored in a lowland British deciduous woodland such as Monks Wood, or Wytham, may be in the soil rather than the trees (although site-specific details matter – see Chapter 9 for a discussion of some of the complications). The soil is largely out of sight, and too often in the past it has also been out of mind to many ecologists; however, as these data show, it is of great importance, to carbon storage as well as to many other ecological processes.

There is another way of looking at carbon storage in woodland. Rather than asking how much is stored altogether one can ask how much is stored – or lost – each year. A financial analogy would be to consider not the total amount of money you have in the bank, but rather how fast your balance is growing (or shrinking) over a year. There are two main ways of measuring this, and both have been applied to Wytham. The first is to census the amount of plant material and measure its growth over time – that's why some of the trees in Wytham have dendrometer bands around their trunks (designed to measure

FIG 42. Old trees in Wytham Woods. A woodland like this is obviously a store of carbon – fixed and turned to wood via photosynthesis – but there may be even more carbon stored in the soil beneath than in the trees and other vegetation.

the expansion of the trunk's girth). The second more high-tech approach is to measure the movement of carbon dioxide between the atmosphere and the woodland vegetation and soils. The standard way of doing this is a method called 'eddy covariance', utilising a set of scientific instruments mounted on a flux tower. The tower is usually built from scaffolding and allows measurements of things such as carbon dioxide, water vapour, wind and light above the woodland canopy; the one in Wytham Woods was installed in 2006. The term 'flux' refers to the fact that the idea is to measure the *movement* of carbon dioxide between the atmosphere and the vegetation under study. As described earlier in this chapter, oxygenic photosynthesis takes carbon dioxide from the air while respiration is in effect the reverse (releasing carbon dioxide).

This method for estimating the intake (or loss) of carbon from a wood such as Wytham effectively looks at the balance between photosynthesis and respiration. If you imagine perfectly still air above a woodland canopy, then the carbon dioxide levels in the air will be affected by the combined effects of photosynthesis and respiration in the woodland below. Of course perfectly still air is extremely unlikely, and the wind will be mixing the air up, especially by causing eddies (more easily seen in flowing water than air). Hence the need for detailed measurements of air movement as well as carbon dioxide concentration, to allow a mathematical analysis of the movement of carbon dioxide between wood and air. Using this approach, measurements at Wytham between 2007 and 2009 showed a net fixation of carbon of 1.2 tonnes per hectare per year, although the authors of the study thought this was probably a conservative value, with the real one being a bit higher (Thomas *et al.* 2011, Fenn *et al.* 2015). They compared this with a related study that used the alternative approach of measuring growth in trees and other plants in Wytham over a similar time span of 2006 to 2008. This gave an estimate of 1.7 tonnes of carbon per hectare entering the woodland system annually. So the two different approaches gave relatively similar values and suggest that the wood is a strong carbon sink. In part this may be because the woodland is regenerating following the cessation of much active management in the mid-twentieth century.

In the context of worries about the climatic effects of increased amounts of carbon dioxide being added to the atmosphere, these results sound encouraging, and certainly illustrate why people worry about the effects of forest clearance and often advocate increasing forest cover. However, calculations suggest that even if we could create extensive forests on a global scale this could only contribute part of what is needed to counteract the predicted human-induced climate changes (Lenton & Vaughan 2009).

CONCLUDING REMARKS

Deciduous woodlands, such as Wytham, illustrate important ideas about photosynthesis and decomposition. These are universal ecological processes (Fig. 43), but the solid nature of a large tree and the piles of dead leaves in autumn make woodlands a particularly good illustration of these fundamental ecological ideas. The solidity of a tree trunk makes the fact that it is primarily made of carbon dioxide from the air a very counterintuitive idea, but that is indeed the nature of photosynthesis. The most important type of photosynthesis on Earth is oxygenic photosynthesis (used by all plants and many photosynthetic microbes too), and this is the main source of the oxygen in our atmosphere. Although large trees are obvious to our eyes, the microscopic photosynthetic plankton in the oceans are very important in global photosynthetic rates and especially the production of atmospheric oxygen. Photosynthetic organisms also form the base of most food chains, channelling solar energy onward to other organisms including herbivores and decomposers. Clearly photosynthesis is of crucial importance for the survival of much of life on Earth.

FIG 43. Although the processes of photosynthesis and decomposition have been discussed in this chapter largely in the context of woodlands, they are of course very widespread ecological processes. Here the dung fungus Egghead Mottlegill *Panaeolus semiovatus* is breaking down cattle faeces in Cwm Idwal, Snowdonia.

The autumnal fall of leaves in a deciduous woodland illustrates the efficiency of decomposition. These leaves do not build up year on year but are broken down, providing a source of energy to decomposers such as many fungi, and releasing nutrients which would otherwise be locked up in great piles of dead organisms. Wood is particularly difficult for many organisms to decompose, and fungi are important decomposers of woody material. Such decaying wood is an important habitat not only for fungi but also for many invertebrates, such as rare beetle species. Invertebrates also play a role in the breakdown of leaf litter, primarily by breaking the plant material up into smaller particles which are then more easily broken down by microorganisms. Woodlands can also form important stores of carbon, which is why deforestation is a concern in the context of climate change. However, somewhat counterintuitively, there is sometimes more carbon stored in the soils of deciduous woodlands than there is in the more visible trees and other vegetation. This carbon is the result of decomposition of previous plants and other organisms extending back hundreds if not thousands of years.

CHAPTER 4

Moor House: Thinking Big While Looking at the Very Small

Upper Teesdale in the middle of a heatwave. I park the car overlooking Cow Green Reservoir. This whole area is a classic location in British natural history, and important personally too for its role in my early career as an ecologist. To the east of where I have parked is Widdybank Fell, famous for its rare arctic–alpine plants. The reservoir in front of me is famous too, as it was created at the start of the 1970s and at the time was a great cause célèbre for British conservationists as it flooded part of the habitat for the rare plants. A few kilometres' walk downstream of here is the photogenic waterfall at High Force. Here is one of the most important sites for Juniper *Juniperus communis* south of the Scottish border, and this was the subject of my Masters degree research in the mid-1980s. Sadly now the Juniper here is struggling because of infections by a 'fungal-like' organism (the oomycete *Phytophthora austrocedrae*), most likely introduced to Britain with exotic tree imports (Green *et al.* 2015).

However, on this occasion I am heading west from Cow Green, upstream, away from the arctic–alpines and High Force in weather that is decidedly untypical of the Northern Pennines. As the morning warms up I find myself shedding clothing and applying more sunscreen, despite this being an environment that more usually favours warm clothes and waterproofs. It's summer 2018 and the drought conditions make the Tees easy to ford without having to remove my boots (Fig. 44). Crossing the river, I enter the National Nature Reserve. I head uphill, passing drying-out patches of bog moss *Sphagnum*, as I head towards the site of an old research station.

This centre, comprising accommodation and laboratory space, was set up at the formation of Moor House National Nature Reserve in 1952; this reserve was

FIG 44. View across the River Tees onto the Moor House reserve. The old research station was in a valley behind and to the left of the woodland. The white dome on the skyline is the radar station on the summit of Great Dun Fell.

later to merge with the Upper Teesdale reserve (which includes the Teesdale rarity sites) to form the largest and highest National Nature Reserve in England. It was this altitude, and the associated climate, that attracted the interests of W. H. Pearsall and Verona Conway when they arranged for a research station to be set up in an old shooting lodge on the Moor House reserve to facilitate ecological research in this upland environment. The station ran for almost three decades, until 1979 (Adamson 2009).

After fording the Tees I head up a low hill past a small conifer plantation and across a moor scattered with various ecological experiments before arriving at the research-station site. The main residential accommodation and most of the laboratory space are sadly long gone (demolished in 1999), but a few small remnants provide some much-reduced support for ongoing research, much of it focused on the potential effects of a changing climate (Fig. 45). At its peak in the late 1960s and early 1970s the research here was part of Britain's contribution to the International Biological Programme (IBP) – an ambitious attempt at the type of big science more often associated with various types of physicists, rather than ecologists. The idea was to make measurements in

FIG 45. The remains of the Moor House research station, photographed in summer 2018. Until 1999 there was an old shooting lodge and some laboratory buildings here.

a range of ecosystems around the world, particularly focusing on ecosystem productivity and function.

Because of its upland climate and vegetation Moor House contributed to the 'tundra' part of the IBP (Fig. 46). Arguably the whole IBP idea was too ambitious for the time, with the technology just not good enough to make much progress on these global ecosystem problems. Computing power was then very limited, making it hard to do much with large data sets – for example, the fastest computers in the world in the late 1960s (not available to most ecologists!) were much less powerful than a top-of-the-range desktop computer in the mid-1980s when I was a Masters student working in Teesdale, and these 1980s machines seem impossibly slow by today's standards.

Even more relevant to the topic of this chapter is the fact that the application of molecular biology to ecology hardly existed at the time. When the IBP work started it was only 16 years after Watson and Crick had published their famous structure for DNA. The central topic of this chapter is the importance of very small organisms in ecology. Today we regularly use DNA (and the related RNA) to study such organisms in soil and water samples. We

FIG 46. Cow Green Reservoir in winter. The Moor House reserve runs to the north (right) of the reservoir. In cold winter conditions one can see why this area was included under 'tundra' in the IBP. The hill behind the car park at Cow Green is used by locals for skiing in winter.

can identify them by recognising their genes in a sample of soil or water, rather than having to find and identify the organisms themselves under a microscope. However, at the time of the IBP, studies of the very small relied on microscopy, and a few biochemical methods – many of which look extremely primitive by modern standards. So the technology greatly limited our ability to study the ecology of microbes. This is nicely illustrated by our current knowledge of the amoeba-like protozoa (amoeboid protists): almost all of the species described by science are from just three groups of unusually large amoebae, because the smaller species were extremely difficult to study even with good microscopes (Lara *et al.* 2020). Molecular methods are now allowing people to start to catalogue the diverse array of smaller protozoa. There is a really exciting consequence following from all this, for technological advances of the last few decades mean that microbial ecology is now a rapidly expanding area of research – ripe for new discoveries. This chapter gives an overview of this rapidly expanding area of ecology.

IT'S A SMALL WORLD

The fact that Moor House played a role in the large-scale science of the IBP is interesting, as much of the work carried out here was on rather small creatures – often overlooked but crucial to global ecological processes. These studies included those by Bill Hale on springtails, Bill Block on mites, Ian Hodkinson looking at jumping plant lice (psyllids), and Val Standen working on enchytraeid worms (which tend to replace earthworms in peaty soils). These small animals, verging on the microscopic, can be present in extraordinary numbers. For example, at Moor House there can be over 200,000 enchytraeids per square metre, with 80,000 springtails and 30,000 mites. Nematodes can be even more abundant – with perhaps 3 million in soil from beneath a square metre of surface vegetation (Coulson & Findlay 2018). Much of this work was based at the University of Durham, and in the early years it was mainly supervised by Jim Cragg, then Professor of Zoology at the university, who was largely responsible for the focus on invertebrates and their role in how ecosystems function.

The fact that much of the animal diversity around Moor House is very small makes sense, as in a hard winter such tiny animals can survive snugly under a blanket of insulating snow. The same is true of Britain's largest expanse of tundra-like habitat, the Cairngorm plateau (described in much more detail in Chapter 7). Look at the contents of patches of moss from the high summits of the Cairngorms with a microscope and you can find a wide range of small animals and microbes very similar to those studied at Moor House, but larger animals are much rarer, especially in winter (Figs. 47 & 48).

However, the importance of a snow blanket for a soil's microclimate was brought home to me most dramatically not at Moor House, or in the Cairngorms, but while helping colleagues sample a bog surface under a metre or so of snow at 1,880 m in the Alps in winter. Despite the sub-zero air temperature, below the snow the surface of the mountain peat bog we were sampling was just above zero and still wet with liquid water (Fig. 49). Such situations are common in mountain and polar habitats, and indeed warming through climate change may counterintuitively make winter conditions colder and more extreme for many small organisms if the warming reduces the insulating blanket of snow.

FIG 47. The southern Cairngorm plateau in near whiteout conditions. The microclimate for small organisms under the snow tends to be warmer and less variable than that experienced by mountaineers and other large organisms.

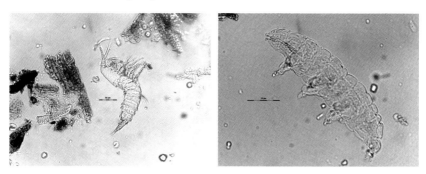

FIG 48. Small animals from moss samples from summits in the Cairngorms (collected from next to monitoring plots set up by the GLORIA project to study long-term changes in mountain vegetation). In both cases the scale bar is 50 μm = 0.05 mm long. Left: a harpacticoid copepod, living in the water film on the moss. Right: a tardigrade, or more informally a 'water bear'. Its mouth and buccal tube can be seen in its head (to the left in the photo), leading to the less distinct gut. There are at least 68 terrestrial species of tardigrade living in Britain, but this is probably an underestimate given that it's a poorly studied group of animals (Flemming 2008).

FIG 49. The author digging down to the surface of a peat bog in the Swiss Alps in February 2009, a photograph taken by my colleague Edward Mitchell on my camera while he had a rest from digging. The air temperature was sub-zero but under the snow the bog had liquid water at a temperature of 0.1 °C. Although at a much greater altitude than other bogs in this study, the greater depth of snow at this site meant that organisms living in the bog experienced winter temperatures hardly lower than at sites more than 1,000 m lower in altitude (Lamentowicz *et al.* 2013).

Changes in the extent of snow cover thus form an important research topic at high altitudes and latitudes (Convey *et al.* 2018).

The copepod and tardigrade shown in Figure 48 may seem small compared to the organisms most naturalists look at, but by the standards of most life on Earth they are enormous. Biodiversity is dominated by single-celled microbes, something that's not at all obvious in the TV nature documentaries, which are dominated by very large animals.

Some of the Moor House studies looked at microbes too. One of the early PhD studies based partly at the field station in the late 1950s was by O. W. (Bill) Heal on testate amoebae living in bog moss (mainly the Flat-topped Bog Moss *Sphagnum fallax*). Testate amoebae are an informal grouping of single-celled amoebae that build themselves shells to live in; they are common in many mosses, leaf litter, soils and fresh water (Fig. 50). Mossy peatland sites can be especially rich in these protozoa. My term 'informal grouping' signifies that although the various species seem to have a lot in common, testate amoebae actually include several types of protozoa that are only very distantly related to each other. At the time of Heal's PhD work, not much was known about their ecology, especially in Britain where little beyond a few species lists was available.

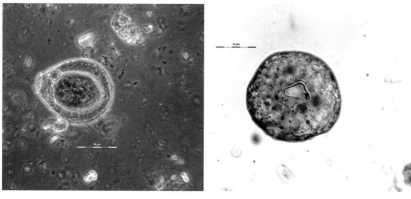

FIG 50. Testate amoebae from moss samples from summits in the Cairngorms (as with Fig. 48, collected from next to monitoring plots set up by the GLORIA project to study long-term changes in mountain vegetation). In both cases the scale bar is 50 μm = 0.05 mm long. Left: *Nebela tincta*, with the body of the organism visible inside its shell. To feed, it projects part of its cell – pseudopodia – through the mouth at the left-hand end of the shell to capture small food particles. This was one of the commoner species of testate amoebae found at Moor House by Bill Heal during his PhD studies. Right: the shell of *Trigonopyxis arcula*, with the triangular mouth in the shell clearly visible. This is another species that Heal found at Moor House.

From his doctoral work he produced a number of research papers based wholly or partly on data from Moor House (e.g. Heal 1962, 1964). However, few people in Britain followed up this research at the time, and this protozoan work had little impact on the IBP studies – which mainly focused on bacteria and fungi in soils when it considered microbes. When I started working on testate amoebae in the 1980s very few scientists working in Britain studied them. However, Heal's work was unusually ecological for the time and there is now a much greater interest in research of this type. For example, when Mariusz Lamentowicz and colleagues were looking for data on how testate amoeba populations change during the course of a year in European upland sites to compare with their data from a bog near the Eiger in Switzerland, Heal's 1964 paper in the *Journal of Animal Ecology* was one of the few relevant data sets they could find. This is the site with liquid water beneath the snow described in the previous paragraph and shown in Figure 49, one of the more spectacular I have worked on, with a view dominated by the North Wall of the Eiger.

Describing the natural history of microorganisms is much harder than writing about plants and animals, because the size the organisms means that they are unfamiliar to most people. The difficulty is compounded by the use of molecular methods (that is, looking directly at genetic material, be it DNA or the related RNA) in recent years, which has greatly changed how we classify microbes and has introduced a whole new lexicon for describing the main groups that would be unfamiliar to someone who did a biology degree only a decade or two ago. Indeed, our knowledge is still changing at a very rapid rate. This is illustrated by the fact that between writing the first draft of this chapter (in 2019) and revising the final version (in 2020) I have had to make assorted tweaks to my description of microbial classification based on new research (e.g. Burki *et al.* 2020). Given such a fast-changing state of knowledge, this chapter has a slightly different style from much of the rest of the book – with greater emphasis on the story of how we have arrived at our currently changing view of microbial ecology. For a hypothetical reader a few years after the publication of this book, some of the details will likely have changed, but this chapter should still give an overview of how we have approached our current understanding.

THE DIVERSITY OF MICROBES AND OF LIFE ITSELF

The heat on that June day at Moor House provided constant encouragement to stop, sit in the heather, drink from my water bottle and watch the world go by. Superficially, the natural history around me seemed to fall into two obvious groups – the plants forming a comfy bed for me to recline on, and the animals such as bees coming to the nearby flowers or the distant waders I was watching through my binoculars. Indeed, historically the most obvious division of life was between plants and animals, at least until we had microscopes, and a wider range of strange-looking tiny organisms swam into the gaze of naturalists. So once microscopes started to be developed in the seventeenth century things started to become a bit more complex.

Early microscopists, such as Antoni van Leeuwenhoek in Delft and Robert Hooke in London, started to use this new technology and found a wide range of small organisms invisible to the unaided eye living in water and other samples. They struggled to make sense of the unfamiliar things they were seeing (Fara 2009). What were they, and how could you be sure that some of what you were looking at was not an artefact of the newfangled optics you were viewing them through? A key date is 24 April 1676, when van Leeuwenhoek first saw very tiny organisms that we now realise must have been bacteria. This was an extraordinary achievement, for his microscope had only a single tiny lens – more high-powered magnifying glass than microscope – but it was so well made that tiny bacteria only a few micrometres long could be seen (something that size would be a barely visible speck in the testate amoebae photos shown in Figure 50). This discovery is also a tribute to van Leeuwenhoek's eyesight. Hooke struggled to use microscopes of this type, finding them 'offensive to my eye', and mainly used two- and three-lens versions which at the time tended to give a less clear image but with less eye strain (Gest 2004).

Jump forward three centuries, and by the time I was doing a biology degree in the early 1980s the consensus was that there were two main types of cell in biology. The eukaryote cell, which contains a wide range of small structures called organelles, as they were seen as analogous to organs in an animal's body, and prokaryote cells, which were simpler and missing these distinctive organelles. Bacteria of various sorts comprised the prokaryotes, and everything else was eukaryotic. This view was changing while I was an undergraduate, although the textbooks at the time hadn't yet caught up with these new ideas. In 1977 Carl Woese and George Fox caused a stir by suggesting that prokaryotes were actually composed of two very different types of microbe, the true bacteria and what have since come to be called the archaea. This latter group is genetically and physiologically very different from bacteria but look confusingly similar under the microscope, so this difference had gone largely unnoticed until technological improvements in the late 1970s allowed more detailed studies to be carried out.

Almost 40 years ago my main general biology textbook as an undergraduate was the third edition of William Keeton's book *Biological Science* (1980). This described eukaryotes and prokaryotes (but without archaea), and it also said that the great majority of organisms were eukaryotes. We now know this to be completely wrong, with molecular methods showing that if you take a gene's eye view of the diversity of life almost all of the diversity is prokaryotic (dominated by true bacteria, rather than archaea), with eukaryotes forming no more than a small twig on the 'tree of life' (Hug *et al.* 2016). The same is true if you take a biochemist's eye view – the diversity of chemical reactions that prokaryotes can perform being greater than that carried out by eukaryotes. This point is extraordinary, counterintuitive and worth emphasising: all of the biodiversity we are used to seeing, the full range of larger fungi, plants and animals, is a vanishingly small sample of the genetic diversity of life on Earth, which is predominantly bacterial. This is a key message of this chapter.

Naturalists in the field tend to overlook bacteria, as they are too small to see with the naked eye or a hand lens – and most bacteria still look extremely small when viewed with a conventional compound microscope with magnifications of a few hundred times up to around ×1,000. However, when present in large numbers you can sometimes see them without magnification. For example, a bloom of cyanobacteria (once called blue-green algae, before their prokaryote status was appreciated) can colour lakes a greenish blue (see Chapter 5). The effects of bacteria also become visible when they play a role in precipitating iron oxides in water draining from coal-mining operations to create spectacular orange pollution – so-called acid mine drainage (Fig. 51). This creates extreme conditions ideally suited to some kinds of archaea, and for many years archaea were thought to be mainly specialist microbes of extreme environments, before we started finding many in less exotic habitats too.

Sometimes you can actually see bacteria if they form colonies large enough to be visible with the naked eye, as with the spherical colonies of the cyanobacterium *Nostoc* that can sometimes be found in small pools on limestone pavements (Fig. 52). However, because eukaryotic microbes are more visible under the microscopes used by many naturalists, and often very beautiful, they feature more prominently than bacteria in the examples used in this book. If genetic diversity were a guide to what to illustrate, almost all of the illustrations in the book would be bacterial, assuming that we consider a virus not to be fully alive and hence not part of biodiversity. Indeed, the status of viruses is an interesting question. Some scientists would say that they are not alive because they cannot reproduce on their own but must use machinery from their host to help them reproduce. This is not clear-cut, because actually defining what it is to be alive is very tricky, and reproduction doesn't feel like the full story. For example, a mule is the offspring of a female horse and male donkey; mules are almost always

FIG 51. The characteristic orange-stained waters associated with acid mine drainage in Worsley, Manchester. The canal was built in the eighteenth century for transporting coal from the Duke of Bridgewater's coal mines – with extensive underground canals to access the coal deposits. The colour comes from the oxidation of iron sulphides in the rock, which can lead to highly acidic water. These chemical processes are often speeded up by the action of prokaryotic microorganisms (Madigan *et al.* 2012). The waters in the canal used to be even more spectacularly orange, until water treatment to reduce the problem was introduced in 2004 – although, as this 2018 photograph shows, this has not been entirely successful, at least in part because passing boats disturb iron-rich sediments on the bottom of the canal.

infertile, but we would consider them to be alive. Alive or not, viruses are of undoubted ecological importance as parasites of undisputedly living things, and are discussed in this context in Chapters 5 and 8.

Over recent decades our understanding of the microbial nature of most of biodiversity has increased enormously, and there is space here for no more than a brief summary. An accessible account of how our appreciation of microbial diversity and the tree of life has developed is given by Quammen (2018), while a more technical account by a historian of biology can be found in the excellent book by Sapp (2009). As freshwater microbes feature prominently in the next chapter,

FIG 52. When bacteria, in this case photosynthetic cyanobacteria, form colonies they can become visible to the naked eye. *Nostoc* 'pearls' in a pool on a southern Cumbrian limestone pavement. The coin used for scale has a diameter of 30 mm.

the rest of this chapter concentrates on eukaryotic soil microbes (such as the testate amoebae studied by Bill Heal at Moor House) to illustrate microbial diversity.

THE DIVERSITY OF EUKARYOTIC LIFE

Traditionally, living things were classified as members of two kingdoms, plants and animals. But suggestions that this simple plant/animal split couldn't cope with the complexity we see in the world go back to the nineteenth century. Richard Owen, a brilliant anatomist sadly now best known for being on the wrong side in arguments about Darwinian evolution, proposed a third kingdom, 'Protozoa', for the wide range of microbes that seemed neither plant nor animal (Sapp 2009). This was an important achievement, albeit largely overlooked by his vertebrate-obsessed biographers, and during the late nineteenth and the first part of the twentieth century a number of other people also suggested a need for more than two kingdoms. Despite these early intimations that the plant/animal dichotomy was inadequate to encompass all of life, it was not until the 1960s that an alternative took hold in most textbooks. By the time I was an undergraduate in the early 1980s a five-kingdom classification had become common, though in the main general biology textbook I used (Keeton 1980) the use of the five-kingdom approach was new to that edition.

The five-kingdom classification is especially relevant to a book on ecological ideas, as it was first developed by the plant ecologist Robert Whittaker at Cornell University in New York. He took an ecological approach to his classification, starting originally with a three-kingdom classification of producers (plants), consumers (animals) and decomposers (microbes and fungi). He subsequently split his decomposer kingdom, separating the fungi from the rest, and then subdividing the microbes into bacteria and eukaryotes, publishing this five-kingdom scheme in a paper in *Science* in 1969. Subsequently the idea was taken up by the microbiologist Lynn Margulis (Fig. 53), who added an evolutionary spin on the classification and vigorously championed the approach in her writings (Quammen 2018). So eukaryotic life was split between four of the five kingdoms, encompassing plants, animals, fungi and then everything else lumped into a large

FIG 53. Lynn Margulis, photographed in 2009 at James Lovelock's 90th birthday party.

mainly microbial group called protists (although Margulis, who seemed to thrive on being heterodox, preferred the term protoctista).

This protist kingdom looked problematic to many biologists, as it was mainly defined negatively – any eukaryote that wasn't a plant, animal or fungus was classed as a protist. The result was that this kingdom contained a wide range of different organisms, mainly microbial but also including large algae such as seaweeds. More recent molecular studies have confirmed what many suspected, namely that protists are not all closely related to each other. For example, animals and fungi are much more closely related than many groups of protists (Fig. 54). In addition, the discovery of archaea in the late 1970s makes a single kingdom for all prokaryotes look problematic too. Eventually even Lynn Margulis accepted that this complicated the five-kingdom classification, and the final edition of her book *Five Kingdoms* changed its name to *Kingdoms and Domains*, although within

FIG 54. Lichens and seaweed on a rocky shore on the Inner Hebridean island of Mull. Both the fungi that make up the bulk of the lichen's body and the seaweeds used to be classed as 'plants'. In fact the fungi are much more closely related to the photographer than they are to plants, while the brown seaweeds are only very distant relatives of either plants or animals, and are closer to microscopic diatoms like the one shown in Figure 59. Other protists, such as the protozoan *Amoeba* commonly illustrated in school science textbooks, are closer to animals than to either brown algae or diatoms.

the book she still largely organised life around the concept of five kingdoms (Margulis & Chapman 2009).

A more recent attempt to classify all eukaryotes, developed by scientists with an interest in eukaryotic microbes, has split them into so-called 'supergroups'. There are seven of these currently suggested, along with a collection of organisms not yet studied in enough detail to be allocated to any group with confidence. These groups don't particularly match the divisions of the old five-kingdom approach: for example, animals and fungi, and according to some researchers some of the testate amoebae, are placed in a single supergroup (Burki *et al.* 2020).

The people who have developed these ideas have deliberately avoided suggesting exactly how this classification fits into more traditional classificatory categories such as kingdoms or phyla. If the exact evolutionary relationships between the different organisms really matter to the questions you are interested in, then there is no avoiding these complex classifications. However, if your main aim is to get a feel for biodiversity, and you have a particular interest in the larger non-microscopic organisms, then something like the five-kingdom classification can still be a useful starting point, making it clear that there is much more to life on Earth than just plants and animals. The key ecological point is that almost all of diversity is microbial. The organisms a naturalist can see without resorting to a microscope are a tiny fraction of eukaryotic diversity, which is in turn a small fraction of the total diversity of life.

PROTIST DIVERSITY IN SOILS

To get a feel for the diversity and importance of protists in soil we leave the high-altitude peaty soils of Moor House for the better-studied soils of the agricultural lowlands. A good place to start is a rather special field in Hertfordshire. The field is the Broadbalk experiment at Rothamsted (Fig. 55; the background to research at Rothamsted is described in more detail in Chapter 11). Wheat has been grown here continually since 1843 under a range of different fertiliser regimes, making it one of the longest-running agricultural experiments in the world.

When the Broadbalk experiment was started in the mid-nineteenth century it was widely thought that protists (often called infusoria at the time) were exclusively aquatic organisms and only found in soils when accidentally washed in. The reason that they got the name 'infusoria' is that a good way to find such organisms, ever since the studies by van Leeuwenhoek, was to make an infusion of organic matter, such as hay, in water and then after a few days look at the

FIG 55. The Broadbalk Winter Wheat Experiment at Rothamsted Research, one of the first places where the important ecological role of soil protists was identified. The small wood, behind the large tree in the foreground, is Broadbalk Wilderness, discussed in Chapter 9.

result under the microscope. As explained in a nineteenth-century encyclopaedia of microscopy (Griffiths & Henfrey 1875; see also Fig. 56):

> *Every one who has examined with a microscope a drop of water containing animal or vegetable matter which has been set aside for a time, or a drop from any pool or ditch, must have observed numerous minute beings in active motion.*

A good example of the historic lack of understanding of soil protists is provided by the ciliate protist *Colpoda*: we now know that it is extremely common in soils worldwide, but in the nineteenth century it was viewed as just an aquatic species (Wilkinson *et al.* 2012a). Rothamsted in general, and Broadbalk in particular, is where these views started to change.

From 1910 onwards scientists studying the microbes in the Broadbalk soil started to be interested in the protozoa as well as the bacteria. This followed experiments at Rothamsted where soil samples were partially sterilised by John Russell and colleagues, who found that this led to improved plant growth. Their

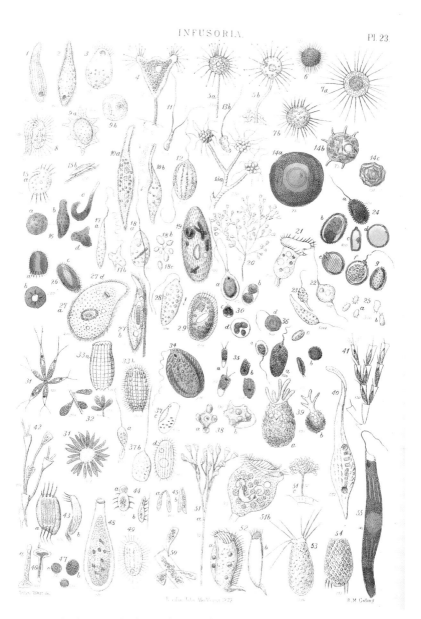

INFUSORIA.   Pl. 23

FIG 56. The diversity of infusoria from Griffiths & Henfrey (1875). This plate contains illustrations of examples from the four classical protozoan groups: ciliates, flagellates, naked amoebae and testate amoebae.

interpretation at the time was that they had killed the protozoa but left some of the bacteria alive. They thought this increased plant growth because many of the bacteria were beneficial to the plants and had hitherto been kept at low population levels by the predatory protozoa. So it seemed that protozoa were not just accidentally washed into soils but an important part of soil ecology as predators of other microbes (Wilkinson *et al.* 2012a).

This story provides an interesting example of the complexity of understanding ecology, or indeed any other area of science. The initial reason that the soils were only *partly* sterilised was due to a laboratory error, 'which caused some little annoyance at the time' (Russell 1914). However, this turned out to be a rather fruitful accident, and Russell realised the importance of the unintended experiment. The conclusion he drew, that protozoa were important in soils, is indeed correct, although it was very controversial at the time. The role of protists, including the predatory protozoa, in soils is now a very active area of research, but Russell was wrong in his explanation for why the partial sterilisation increased crop growth. We now know that predation by protozoa on bacteria tends to release lots of nutrients in soils and thus in general increases crop yield, so killing protozoa should in theory have decreased crop growth (Gao *et al.* 2019). With the benefit of hindsight, we can say that the increase in plant growth seen by Russell was probably due to an initial burst of nutrients released from the dead protozoa, and many other microbes, killed during the experimental treatment. But it is interesting to note that a whole new area of soil ecology started at Rothamsted, triggered by a botched experiment and based on a misunderstanding of what the results meant!

It wasn't just in Britain that agricultural research led to the founding of soil protist ecology; for example, something similar happened in Russia during the 1920s and 1930s, again focusing on the interactions between bacteria and protozoa in agricultural soils (Foissner & Wanner 1995). Although the density of individual protists in soil is very variable, we now know that typically it ranges between 10,000 and 100,000,000 per gram of soil. However, even after the early twentieth-century realisation that protozoa were important in soils, most research still focused on bacteria and fungi. Indeed this is still the case in modern ecology. While writing the first draft of this chapter I attended the 2018 annual meeting of the British Ecological Society, in Birmingham. Research on bacteria and fungi in soils was reasonably well represented, but there was very little on soil protists.

In general, soil ecology research has focused on what are called bottom-up factors. The ecological terms bottom-up and top-down are best understood in the context of a simple food chain, such as the woodland one from the previous chapter:

oak leaf ⟶ winter moth caterpillar ⟶ great tit

A bottom-up effect is one that works from the base of the food chain upwards (e.g. low nutrient quality in leaves affecting caterpillars and so tits), while a top-down effect operates in the other direction (e.g. a change in the numbers of tits affecting caterpillar numbers and so oak leaves). In soils we might have a food chain that looks like this:

plant roots ⟶ bacteria ⟶ amoeba

The bacteria may be feeding on the plant roots themselves, or on chemicals leaking from the roots into the soil. A top-down approach to studying such a system would focus on the role of protozoa, such as amoebae, on the bacterial population (and hence their effects on plants too). Such studies have been rare until recently, but are slowly starting to become more common (Gao *et al.* 2019).

One important reason why predation by protozoa on bacteria and fungi is important is that it leads to nitrogen being released into the soil in forms that can be used by plants and other organisms. There are two main ways in which this happens. First, to gain energy the protozoa 'burn' carbohydrates from their prey, using up carbon and so leaving behind an excess of nitrogen which is excreted (the readers of this book do the same thing, excreting excess nitrogen in their urine). In addition, bacteria and fungi contain more nitrogen (N) in relation to their total carbon (C) content than do protists; these are two of the four chemical elements required in the greatest quantity by life (i.e. carbon, hydrogen, oxygen and nitrogen, with phosphorus and sulphur the next two most common elements in biology). The formal way of describing this is to say that protists have a higher C:N ratio than bacteria and fungi. So if a protozoan is feeding on bacteria it consumes more N than it requires in order to get enough C. The spare nitrogen is also excreted into the soil, a potentially important resource for other organisms. Predation by protozoa therefore increases the turnover of nutrients such as nitrogen, and so stimulates general microbial activity and plant growth. This predation also helps maintain microbial diversity, as some protozoa tend to feed selectively on the dominant bacteria, increasing the proportion of the community made up of the rarer bacteria. They may also lead to a reduction in the density of some of the common parasitic bacteria that cause plant diseases. Because of these various top-down effects there is a growing interest in the potential importance of soil protists in sustainable agriculture, for their possible role in increasing nutrient availability to plants and acting as a natural control on agriculturally harmful bacteria and fungi (Gao *et al.* 2019).

Perhaps a more typical environment for a naturalist is a nature reserve, rather than an agricultural research station such as Rothamsted. One might guess that at an upland site such as Moor House the climate would limit microbial diversity, if only through limiting the types of vegetation that can grow there – but what about more lowland sites? A snapshot of protist diversity in such soils comes from Mere Sands Wood, a Wildlife Trust nature reserve near Southport, north of Liverpool (Figs. 57 & 58). The reserve is on the site of old sand quarries and is a nice mix of freshwater and woodland habitats, with a few more open areas of grasslands and heathland. This mix of habitats all on one site is very convenient both for teaching and for research, and I used it a lot while I was lecturing at Liverpool John Moores University.

To get a feel for the microbial diversity of Mere Sands Wood, Angela Creevy – at the time a Masters student supervised by Jane Fisher and myself – looked at the diversity of both testate amoebae and diatoms in soil samples from the reserve (Creevy *et al.* 2016, Wilkinson *et al.* 2017a). Over a couple of days in September 2011 we collected leaf litter and soil samples from four main habitat types: birch and pine woodland, from under Rhododendron *Rhododendron ponticum* agg., and from grassland. Despite the limited extent of the study, we found 41 species of testate amoebae and 52 diatom species (all new to the reserve's species list). The reason for

FIG 57. Mere Sands Wood is a mix of habitats on the site of old sand pits, including woodland, meadow and small areas of heathland, along with lakes and ditches. In addition to the studies of soil microbes described in this chapter, we also looked at testate amoebae and diatoms in the water bodies (see Wilkinson *et al.* 2017a).

FIG 58. Pine woodland at Mere Sands Wood. Samples from under pine produced 12 species of testate amoebae but no diatoms – presumably because of the low light levels.

concentrating on these two groups of microbes is that they have distinctive 'shells' which allow them to be both identified and counted under a microscope. So, in contrast to many types of microbes, you can attempt to directly count the number of individuals, just as you might count the waterfowl on the lakes or the trees in the woodlands. This gives you data on the microbes which are very similar to those collected by ecologists studying the more familiar larger organisms.

These microbes have a wide range of functions in the ecology of the reserve's soils. One focus of our studies was their potential role in the soil silica cycle (silica is a compound of the chemical element silicon). Both the diatoms and some of the testate amoebae protect themselves within silicon-rich 'shells', so the remains of these microbes potentially create 'hot spots' of silicon in the soils that can be utilised by plants in need of this element. In general, the more species of diatoms and testate amoebae there were in the soil the greater the potential influence on the silica cycle, so microbial biodiversity, as well as the total number of microbes, mattered to their role in soil nutrient cycling. This is interesting in the wider context, as Mere Sands Wood is a reserve set in an intensively farmed landscape, where microbial diversity in the soils is likely to be greatly reduced. It's been suggested that nature reserves can be viewed as important refuges for soil microbial diversity in intensively farmed landscapes, potentially allowing beneficial microbes to recolonise agricultural fields if more microbe-friendly agricultural practices are introduced.

In this context of biodiversity playing an important role in agricultural soils, we more often think of invertebrates such as earthworms, rather than microbes. As Darwin (1881) rather poetically put it some 140 years ago:

The plough is one of the most ancient and most valuable of man's inventions; but long before he existed the land was in fact regularly ploughed, and still continues to be thus ploughed by earth-worms.

The traditional way to think about food chains involving soil invertebrates is that it is photosynthesis by plants that ultimately provides the energy supporting

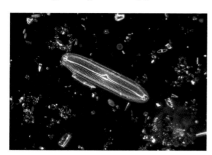

FIG 59. The diatom *Pinnularia viridis* from soils at Mere Sands Wood, found both in moss-rich open vegetation on one of the islands and in a grassy meadow. This species was also found in several of the lakes on the reserve, as well as in one of the drainage ditches. This is a large diatom which can be up to several hundred micrometres (μm) long (100 μm = 0.1 mm). (Angela Creevy)

all the soil animals. In many cases this is probably correct; however, recent research is starting to add microbial complications. Studies on soils collected from spruce woodland and agricultural fields in Ireland have shown that photosynthesis by microorganisms in the soil (such as cyanobacteria, diatoms and green algae) was contributing to the soil food chains, providing some of the energy used by earthworms, springtails and slugs (Schmidt *et al.* 2016). How important this is will vary between different soils. For example, at Mere Sands Wood we found no diatoms in soil under dense pine or Rhododendron – where it is presumably too dark for photosynthesis – but more diversity in open birch woodland and meadows (Fig. 59). It's probably a fair summary to say that the more we know about soil microbes, the more important they seem to be to soil ecology.

LIFE AT THE GLOBAL SCALE

What about life at the global scale? How does our increasing knowledge of microbes alter how we view natural history? The impression given by most TV natural history programmes or the mix of books filed under 'natural history' or 'wildlife' in a large bookshop is that vertebrates dominate, with plants and the more colourful invertebrates featuring – but often as bit-part players. The types of organisms that have dominated this chapter feature hardly at all. So what is life on Earth really like? I have already described how at a genetic level bacteria seem to dominate, and that animals form a vanishingly small proportion of the diversity of life. However, ecology is not just about species lists; the amount (or biomass) of organisms matters too. For example, as described in Chapter 3, plants can play an important role in removing carbon dioxide from the atmosphere, but to have much effect you need plants in quantity. A single tree will have little effect by itself. So how is biomass (the total mass of life) partitioned between the main types of life?

Robert Whittaker – a key architect of the old five-kingdoms classification – was involved with early attempts to answer this in the 1970s, but the available data at the time meant that it was effectively impossible even to guess the biomass of most microbial or fungal groups. We now have much better data – in particular, the rise of molecular methods in microbial ecology means that we can start to address the biomass of microbes in a quantitative way, while remote sensing from satellites has greatly improved data on terrestrial vegetation at large continental scales. Using data from these advances, Yinon Bar-On and colleagues (2018) recently produced the current best guess at the biomass distribution of life on Earth. The summary of their results is shown in Table 1, and the most noticeable thing is that most of global biomass is composed of plants, with bacteria second, while the animals contribute a tiny amount (around 0.4 per cent). However, note the final column in the table giving an estimate of uncertainty. For plants this uncertainty is low, since most of the plants are terrestrial and a combination of ground survey and remote sensing from satellites gives pretty reliable data. However, for bacteria the uncertainty is a factor of 10 – which means it's conceivable that there may actually be a greater biomass of bacteria than of plants. In fact, the bacterial estimate of 70 gigatons of carbon was something of a surprise, as the previous best estimate had been quite a bit higher. Advances in

TABLE 1. Estimated global biomass, based on the work of Yinon Bar-On *et al.* (2018). The mass is given in gigatonnes (i.e. billions of tonnes) of carbon contained in each group. Note the different levels of uncertainty: for example, for bacteria the uncertainty is a factor of 10, so it's possible that the true mass of bacteria could be larger than the global mass of plants. The numbers in the 'Mass' column therefore represent the authors' best-guess estimates. Around half the animal mass was accounted for by marine invertebrates.

Group	Mass (Gt C)	Uncertainty (-fold)
Plants	450	1.2
Bacteria	70	10
Fungi	12	3
Archaea	7	13
Protists	4	4
Animals	2	5
Viruses	0.2	20

technology mean that there are now more (and hopefully better) data to use in this estimate, but clearly there are still substantial uncertainties when we try to 'weigh' the microbes on Earth. However, the idea that the mass of bacteria on the planet is second only to that of plants is extraordinary – bacteria are so small as to be invisible, while plant biomass includes extensive forests of large trees. Another way of looking at this is that it illustrates just how impossibly numerous bacteria are, if their total biomass can be so large while each individual is so tiny.

The sheer size of the bacterial biomass is not the only counterintuitive aspect of bacterial diversity. Bar-On and colleagues estimated that the majority of bacterial and archaeal biomass is deep underground. Indeed, because of this they estimated that around 15 per cent of total biomass on Earth is 'deep subsurface' – that is, living deep in sediments, and in the cracks and pores in rocks. The realisation that a substantial proportion of life on Earth actually lives deep within the Earth is barely 25 years old. When the astrophysicist Thomas Gold started arguing for the importance and extent of what he called the 'deep hot biosphere' in the early 1990s he was met with considerable disbelief. However, since then more extensive data on microbes from deep mines and boreholes have shown that he was correct (although other aspects of his ideas relating to the origin of oil haven't met with the same success). While some of the energy for these microbes may come from surface organic matter (ultimately from photosynthesis) making its way down into the Earth's crust, much of this life appears to be sustained by the breakdown of chemicals in rocks. It's currently not very clear if most of these microbes are isolated from those living on the surface or if many of them have derived from surface-living microbes in the geologically recent past (Colman *et al.* 2017). Certainly many of them are adapted to hot conditions due to heat from the Earth's core – until recently an aspect of the environment familiar to many British coal miners who even in 1980s Yorkshire would sometimes work in just a pair of shorts because of the subterranean heat (Cave 2005). Given the difficulty of studying such organisms, and the fact that we have only a couple of decades' worth of studies, it's clearly possible that the estimate of 15 per cent of global biomass being deep under our feet may need revising (up or down) in the future in the light of new studies.

It is also possible to use these data to examine the distribution of the biomass across different environments. Is most of it found on land, or in the sea, or deep below the surface? Terrestrial systems win, because of the large amount of plant biomass, with the deep biosphere second, well ahead of marine systems (Fig. 60). Freshwater systems don't feature at all at this scale of analysis because of the limited extent of lakes and rivers compared to these other environments. Although there are considerable uncertainties in some of these estimates, because the relative differences are so large the basic conclusion – of terrestrial first and marine last – is likely to be robust. Even though there are such large differences between the biomass of terrestrial and marine systems, primary productivity (amount of energy used by photosynthesis) of the two is roughly equal, since much of the plant biomass on land is non-photosynthetic wood and other structural plant tissue, while in the sea it's mainly microscopic plankton.

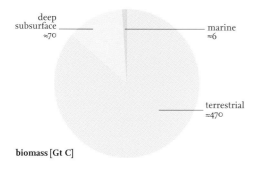

deep subsurface ≈70

marine ≈6

terrestrial ≈470

biomass [Gt C]

FIG 60. Biomass distribution across different environments, showing the domination of terrestrial plants, with the microbes of the deep biosphere coming second, well ahead of marine biomass. Data from Bar-On *et al.* (2018).

CONCLUDING REMARKS

The main message of this chapter is easy to summarise – most of us have no idea of the real nature of biodiversity. At a genetic level, life on Earth is dominated by bacteria, organisms we usually can't even see. I have chosen my words carefully in that last sentence because of one final complication I haven't yet mentioned. Why write 'at a genetic level' rather than just write about 'species richness'? If bacteria are genetically more diverse, doesn't that imply they have more species than the other groups? Perhaps, but there is a problem. For most organisms large enough to be visible we are used to the idea that their genes all come from their parents (technically it's called vertical transmission). In prokaryotes, however, there is extensive horizontal transmission of genes – that is, acquiring some genes from other organisms or even the general environment during an organism's lifetime. This makes it hard to define what a species is in bacteria or archaea, and indeed at one extreme some biologists think that because genes can potentially swap between very distantly related prokaryotes it makes the concept of species only applicable to eukaryotes – although even here there can be some more limited horizontal gene transfer (for a readable introduction to the development of ideas on horizontal gene transfer see Quammen 2018).

Horizontal gene transfer matters to the ecology of bacteria, because it potentially allows them to pick up the genes they need to adapt to their local environment. It matters to humans too, for it's likely that this process will be involved in killing some of the readers of this book, as it has a role in the spread of antibiotic resistance in bacteria. If we look at diversity using biomass, rather than genetics, then terrestrial plants dominate, with bacteria next (or possibly with a similar biomass, given the uncertainty in the data used for these calculations). In neither approach, genetics or biomass, do animals feature near the top of the list – despite many people's assumption that animals make up much of the diversity of life on Earth.

To give this chapter a focus I have concentrated on the diversity and ecology of eukaryotic microorganisms in soils and leaf litter. A bog moss hummock at Moor House National Nature Reserve illustrates many of the points in microcosm (Fig. 61). A great advantage of clumps of moss over soils is that it's much easier to study the microbes under a microscope, as they are not obscured by all the soil particles, and this was one of the reasons why Bill Heal carried out his early work on testate amoebae in mosses at Moor House. Until the early twentieth century the general assumption was that any protists found in soil, moss or leaf litter were of no great ecological importance, having been washed in from nearby aquatic habitats. Although this is wrong, it's true that thin films

FIG 61. Drying bog moss *Sphagnum* hummock during a heatwave on Moor House National Nature Reserve, June 2018.

of water around soil and plant particles are important to these organisms. So as a wet bog moss hummock dries out in a summer drought you might expect that this will affect the protists living in it. Recent studies in Poland and France have looked at the effects of the drying out of bog moss on the microorganisms living in the hummocks (Reczuga *et al.* 2018) – in passing, it's interesting to note that the authors make use of Heal's early work at Moor House in justifying the design of their experiments. They found that the drying out preferentially affected the larger microbes, such as testate amoebae, thus altering the relative proportions of predators (protozoa and microscopic animals) and prey (bacteria and fungi) in the hummock. This decline in predators appeared to affect aspects of the functioning of the bacteria, such as enzyme production. This is no great surprise, for we now realise that the often-overlooked protists are very influential in the ecology of soils and similar systems – the other main theme of this chapter. The role of microbes, and larger organisms, in the functioning of ecosystems is described in the next chapter – using lakes as the main example.

CHAPTER 5

Windermere: An Introduction to the Nature of Ecosystems

S ummer, holiday season in the English Lake District, and the lake front at Bowness-on-Windermere is crowded with international tourists – many from Japan on Beatrix-Potter-inspired holidays. If curiosity has prompted their desire to travel, it clearly doesn't extend to wondering why two middle-aged men are gathering up swan and duck faeces from around their feet (Fig. 62). Perhaps they think we are tidying up for the local council, rather than collecting samples to study the dispersal of plants and small aquatic animals by waterbirds. Beatrix Potter, with her natural history interests, would no doubt have shown more curiosity.

FIG 62. Mute Swans *Cygnus olor* at Bowness-on-Windermere, with Andy Green collecting bird faeces as part of a project on the role of waterbirds in dispersing seeds and small aquatic organisms.

Even without the legacy of Peter Rabbit or William Wordsworth it's easy to see why the Lakeland landscape draws international visitors. One of the Chinese words for landscape is created by the fusion of the words for mountains and water (*shanshui*), suggesting something especially harmonious about this combination, often so intimately entwined in the English Lakes. Windermere itself has been the subject of detailed ecological research since 1929 when the Freshwater Biological Association (FBA) was set up on its eastern shore intending to 'prosecute research into the fundamental problems of freshwater biology' (Toogood *et al.* 2020). Normally I bypass the crowds around Bowness and head for the higher fells, but on this occasion the ready availability of faeces from waterbirds attracted by food from human visitors made the shore at Bowness – and Fell Foot at the southern end of the lake – worth a visit to collect research samples. Indeed, samples we had previously collected at Fell Foot yielded the first ever demonstration that moss fragments can go all the way through a bird's digestive system and still be capable of growth (Wilkinson *et al.* 2017b). Sometimes it's these apparently simple little discoveries that give you the most pleasure. Our study could easily have been done several hundred years ago, as it only needed a sieve, a low-power microscope and a substrate to grow moss fragments on, but pleasingly it had waited for us to do it in the early twenty-first century.

We are hardly the first people to do science here, for the establishment of the FBA labs means that Windermere and surrounding lakes are some of the best-studied lakes in the world (some 600 research papers were published on Windermere and nearby lakes between 1920 and 2006 – with more since). The very first scientist to be based at these labs was a young Penelope Jenkin, who studied both water chemistry and zooplankton in the Lake District, and also carried out freshwater fieldwork in Africa; she later moved away from freshwater research and switched to the study of hormones. Jenkin was the first of an important collection of female scientists to work at these labs; the FBA in the first half of the twentieth century was unusual in this, at a time when women were often treated with prejudice in much of British scientific culture (Toogood *et al.* 2020). The labs attracted a distinguished group, both male and female, and at one point in the 1970s there were four Fellows of the Royal Society working there, including Winifred Pennington, who was using sediment cores to study the past history of these lakes. Much of the longer-term science at Windermere is now carried out by the Centre for Ecology and Hydrology at Lancaster, and for some studies there are now almost 90 years of data (Moss 2015, Maberly *et al.* 2018). The lake must feature highly in any attempt to list the best-studied ecosystems on Earth, and results from studies here will be used in this chapter to introduce the ideas of ecosystem ecology, alongside those from other British lakes – especially Rostherne Mere in Cheshire, another lake with a long history of scientific investigation, including studies by FBA scientists from Windermere.

Given Windermere's importance, a bit more background will be useful. If you take one of the tourist boat trips that run the length of the lake (Fig. 63), starting from the southern end these often stop off at Bowness, midway up, where several islands partially block the long view along the full length of the lake. From here trips continue north to Ambleside at the head of the lake. The boat ride you have taken likely has a long history, as there was a Roman fort on the lake edge at Ambleside, which archaeologists assume was supplied by boat, a much easier way of moving large amounts of material than by road at the time (Shotter 1997). The more recent history of tourism is also important for the ecology of the area. The railway reached Windermere in 1847, and the resulting expansion of tourism and population size in the towns and villages around Windermere and other nearby lakes led to an increase in pollution from sewage and other sources entering the water. The peak of this pollution was around 1940–1980. You can see a visual representation of this pollution marked out in the muds from cores taken from the bottom of the lakes – as pollution increases, the muds turn from brown to black, and then often start to lighten again towards the surface as you move into

FIG 63. Tourist boat on Windermere in the English Lake District. The boat in the photograph is in the southern basin and heading towards the south end of the lake.

muds laid down in the last few decades (Fielding *et al.* 2020). One of the themes of this chapter is how such changes can tell us interesting things about the working of ecosystems.

Although Windermere is listed as the largest lake in England, it is effectively two lakes, the north and south basins being separated by the shallow island-scattered water off Bowness – and indeed in ecological studies these two basins have often been treated as effectively two separate lakes. For example, both basins have shown changes in the timings of peaks of phytoplankton and zooplankton since the 1970s, with populations building earlier in the year – but there are subtle differences between the patterns in the two basins. Tracking these changes in the two basins has helped demonstrate that the plankton populations are driven not only by changes in water temperature, but by changes in nutrient levels too (Maberly *et al.* 2018).

As well as the two basins of Windermere, several other nearby lakes have also been used in the ongoing studies – especially Grasmere, Blelham Tarn, Esthwaite Water and the much smaller Priest Pot at Esthwaite's northern end (Maberly & Elliott 2012). One problem with using Windermere as a research site is that in T. T. Macan's (1970) somewhat exasperated words, 'it has the disadvantage of being most infested with trippers, and therefore the lake on which apparatus is most likely to be interfered with.' He preferred the quieter lakes of Esthwaite and Blelham, with their more restricted access. So although one tends to associate the FBA studies with Windermere – as the largest and best-known lake – it is more accurate to consider this whole collection of lakes in the southern Lake District as a classic site in freshwater ecology. I will use aquatic plants in these lakes and their response to pollution as a way of introducing some of the ideas of ecosystem ecology, after briefly considering why lakes make such good study systems for ecologists interested in ecosystems.

WHY LAKES? A TALE OF TWO CLASSICS

This chapter uses the ecology of lakes as a way of discussing more general ideas about ecosystem ecology. The idea that lakes may be a good way of thinking about such ideas has a long history. In 1887 Stephen Forbes published an essay in an obscure American academic journal on 'The lake as a microcosm', and although many of his ideas are now seen as mistaken, this is often viewed as a classic paper because of its influence on the development of the subject after it was reprinted in a more accessible location in 1925. In this essay Forbes suggested that lakes made a good target for study because they had clear boundaries, and being largely isolated from the surrounding dry land (except when they flooded) simplified the factors you needed to measure. Effectively lakes could be seen as 'islands' of aquatic habitat surrounded by a 'sea' of dry land. We now know that this is too much of a simplification, and that lakes are heavily influenced by things happening in their catchment and indeed more widely. This is a theme that will be returned to later in this chapter.

Forbes had been very influenced by Darwin's ideas on evolution, and the final paragraph of his essay is effectively a description of Darwin's entangled bank (described in Chapter 1) rewritten to describe a lake rather than a flowery bank in southern England (Hutchinson 1964). He saw lakes as relatively stable systems which had similar numbers of organisms living in them from year to year, and thought that natural selection had acted on the inhabitants of a lake to allow them to evolve a system in 'harmonious balance' where no species became too dominant, so damaging the interests of other species and ultimately threatening its own long-term survival by throwing the system out of kilter. Again we now think this idea is wrong, as during the 1960s biologists showed that such simple 'for the good of the species' explanations of evolution don't work (an argument popularised by Richard Dawkins in his book *The Selfish Gene* and discussed in more detail in Chapter 7 of this book).

However, systems such as lakes do show some stability from year to year, and Forbes was correct that this requires some explanation. I will return to this at the end of this chapter and sketch out some recent attempts to explain how some sort of stability could evolve. In his tendency to view both the biotic and abiotic components of a lake as a single entity, his essay foreshadows ideas of the ecosystem which were to develop during the first part of the twentieth century (Jenkins 2014).

A second highly influential theory paper based on lakes came out in 1942 and shared Forbes' view that lakes were useful to ecologists because they were reasonably isolated from the complexities of the rest of the world. It was written by Raymond Lindeman, a young scientist working with the ecologist Evelyn Hutchinson at Yale in the USA – the paper came out posthumously, as Lindeman died while still in his twenties. His approach was new and controversial, and Hutchinson had to lobby hard to get Lindeman's work accepted for publication. The key idea was to focus on the flow of energy through the lake – solar energy being fixed by photosynthetic phytoplankton, which were then eaten by zooplankton, which then fed larger organisms. Crucially, he also emphasised that the biotic components could not be clearly differentiated from the physics and chemistry of the lake. That is, he thought that the whole lake needed to be considered as a system, an 'ecosystem' in the terminology of the British ecologist

Arthur Tansley. As Tansley (1935) wrote, in a now classic paper which influenced both Lindeman and Hutchinson, what is important is

> the whole system (in the sense of physics), including not only the organism-complex, but also the whole complex of the physical factors forming what we call the environment … It is the systems so formed which, from the point of view of the ecologist, are the basic units of nature.

Since Forbes, Lindeman and Hutchinson there has been a tendency for many freshwater ecologists to take a rather holistic view of lakes – 'from physics to fish' is the neat summary of this approach. This idea of ecosystem – with interacting biotic and abiotic components – contrasts with the concept of an ecological community, which is just the set of organisms present at a location and excludes the wider interactions with the rest of the environment. For example, one can talk about the fish community in a particular lake ecosystem while ignoring the wider interactions upon which an ecosystem ecologist would tend to focus.

AQUATIC PLANTS AND POLLUTION IN THE SOUTHERN LAKE DISTRICT

Photosynthesis in lakes can be carried out either by microscopic photosynthetic plankton (such as the diatoms described in the last chapter) or by larger, visible, plants. To differentiate the two, freshwater ecologists tend to talk about phytoplankton (microscopic) and macrophytes (larger plants). Submerged macrophytes are restricted to shallow lakes or the edges of deeper lakes, as they obviously require light to reach them to be able to photosynthesise. This makes water quality important. The simple approach to measuring water transparency is to use a Secchi disc, which is lowered into the water until the point where it is lost from sight, at which point the depth is recorded (Fig. 64). Although light availability for photosynthesis is best measured with more complex equipment, as a rough guide the Secchi depth is approximately 40 per cent of the depth of the euphotic zone, beyond which light becomes too limited for photosynthesis (Moss 2017).

Because Secchi discs are a low-tech approach they have been used for a long time, and so historical data are sometimes available to compare with current values. For the Lake District there are data going back to studies by W. H. Pearsall in the early 1920s, and although he used a smaller disc than is now typical, his results can still be usefully compared with more systematic measurements made over the last 60 years. For example, a recent annual mean Secchi depth for the

FIG 64. Secchi disc in action, being lifted out of the water. These discs are used to make a simple measurement of water transparency – being lowered into the water until the disc is lost from sight, the so-called Secchi depth, which is recorded as a standard measure of water transparency. This is not a British site but Shidou Reservoir in southeastern subtropical China. The photograph was taken during a cyanobacterial bloom in October 2015. High nutrient levels from pollution combine with warm temperatures to make such reservoirs susceptible to cyanobacterial blooms, something which is likely to get worse with climate change. On the day the photograph was taken the Secchi depth was around 1 m; when there is no bloom it can be around 2.4 m in this reservoir, and at the peak of a bloom it can be as low as 0.3 m (Liu *et al.* 2019).

south basin of Windermere was 3.8 m, whereas Pearsall gave a value of 5.5 m for this lake.[1] Better data, allowing more rigorous comparisons, tend to exist from around 1960, as described below for Esthwaite Water.

Submerged macrophytes have declined in many British lakes, a trend of conservation concern which also illuminates a number of interesting ideas about the functioning of ecosystems. An example is the rare aquatic annual plant Slender Naiad *Najas flexilis* in Esthwaite Water. Its population in the lake declined over the twentieth century, and it is now probably extinct. Prior to around 1915 Slender Naiad was reasonably common in the lake, as shown by an early twentieth-century survey by W. H. Pearsall. In shallower waters it was growing with quillworts *Isoetes* and Shoreweed *Littorella uniflora*, and in deeper waters with stonewort algae *Chara*. From around the mid-1940s we have quite good environmental data from Esthwaite Water, mainly from work by FBA scientists. Ideally we would also have regular surveys of the naiad populations, but these are hard to do for underwater plants and so we are short of good data on the plant population.

An alternative approach taken by Isabel Bishop and colleagues (2019) is to reconstruct population changes using plant remains preserved in sediments at the bottom of the lake. This is an approach to freshwater ecology with a long history in the Lake District, starting with work by Winifred Pennington – who based much of her work at the FBA at Windermere in the late 1930s and 1940s – and later continued by others, especially Elizabeth Haworth. Lakes with sediments such as muds underlying them have the great advantage of recording their own history. Take a core of such mud and you can extract the lake's environmental history from the layers of sediments by identifying preserved remains of plants (such as pollen or seeds), the cases of protists such as diatoms or testate amoebae, or the remains of invertebrates such as midge larvae (Chironomidae). In addition, the physical and chemical nature of the muds can tell you much about past history, such as changes in soil erosion in the surrounding catchment, or the history of pollution entering the lake. Peat bogs share this time-capsule character with lakes, preserving similar evidence of environmental history in the build-up of peat (see Chapter 12).

Plant remains from the Esthwaite Water cores showed that the naiad was reasonably common up to around 1915 but then declined over the next seven decades, becoming largely absent from the cores in the early 1980s (the last survey record of the plant in the lake was from 1982), with possibly a few plants in 2002, although these could potentially be seeds that were not fresh when buried in the

---

1 I calculated this modern value from data in Woolway *et al.* 2015; however, other studies from the last 20 years give similar values. The early studies are usefully summarised in Macan 1970.

sediments. Water chemistry records show the lake becoming increasingly nutrient-rich over this time, and the related algal growth reduced light penetration into the lake – for example, the annual mean Secchi depth in 1960 was 4.45 m, while it had dropped to 1.86 m by 1988. The interpretation by Bishop and colleagues was that up to the early twentieth century the lake had mildly alkaline waters with a medium level of nutrients, which suited the naiad. As nutrient levels increased because of pollution during the twentieth century, rates of photosynthesis by other macrophytes and algae in the lake increased, reducing carbon dioxide levels for use by the naiad. In the late twentieth century nutrient levels increased even more, reducing carbon dioxide levels further and also reducing light levels in deeper water as algae increased. This finally caused the extinction of the naiad in the lake. Increased nutrients from human activity in the lake catchment (such as sewage, runoff from agricultural fertilisers and a fish farm which opened in 1981) changed both the biotic and abiotic aspects of the lake's environment.

Better treatment of sewage and removal of the fish-farm cages have improved the situation, and Esthwaite Water is now less nutrient-rich than it was in the 1980s – however, this came too late to save the naiad. This not only provides a nice case study of pollution and the extinction of an aquatic plant species, but illustrates the interaction of nutrients (from pollution in this case), light and competition between different plant species. However, large lakes such as Windermere and Esthwaite Water are rather unwieldy systems to use in starting to think about these ideas in detail, so I move from the Lake District to Snowdonia and hike into the hills to consider some smaller lakes less affected by pollution – where it will hopefully be easier to visualise the key ideas.

SUBMERGED MACROPHYTES – WHAT DO THEY TELL US ABOUT LAKE ECOSYSTEMS?

On a morning in early summer I have just arrived at the edge of Llyn Cau, hot and sweaty from the walk up the Minffordd Path on the side of Cadair Idris. Cwm Cau is a classic Snowdonian mountain cwm (the Welsh term for a glacial cirque or corrie), with the crags at the back a good site for arctic–alpine plants – similar to Cwm Idwal, described in Chapter 2, but more of a walk from the road. The water in the lake is very clear, the surrounding rocks are hard and so only relatively small quantities of nutrients are added to the lake as they slowly weather (Fig. 65). With more nutrients there would likely be more plankton and less water clarity. This clear water is a nice illustration of why lakes can't be considered microcosms or islands – the nature of the lake depends on the

FIG 65. Llyn Cau, on the eastern flank of Cadair Idris in Snowdonia. Note the clear water, which lets plenty of light reach plants growing in the shallower waters around the water's edge (see Figs. 66 & 67).

nature of the surrounding catchment. The clear water also allows you to see the plants growing amongst the rocks in the shallows. I have a waterproof camera with me, and the hot day helps counteract the cold water of the Llyn, allowing my botanising and photography to extend into water a metre or two deep, revealing a variety of macrophytes typical of such lakes. The largest plants are the water-starworts growing up into the water body (Fig. 66), while the rosettes of Water Lobelia and quillwort carpet parts of the lake bed (Fig. 67).

FIG 66. Intermediate Water-starwort *Callitriche brutia* growing in around 1 m of water in Llyn Cau.

FIG 67. The isoetid flora growing in around 2 m of water in Llyn Cau. The rosettes are a mix of Water Lobelia *Lobelia dortmanna* and quillwort *Isoetes* sp.

In starting to think about the working of ecosystems, photosynthesis is clearly rather important, as it's the main source of energy for most systems. In Llyn Cau the beds of rosette-like plants are of particular interest. There are a number of confusingly similar aquatic plants in Britain found in nutrient-poor lakes with this rosette-like growth form – including quillworts, Water Lobelia, Pipewort *Erocaulon aquaticum* and Shoreweed. These, and a few other species, use what is called the isoetid strategy to photosynthesise – named after the quillworts, which are in the genus *Isoetes*. Carbon dioxide is a key raw material for photosynthesis (see Chapter 3), and it can be harder to access for water plants, compared to land plants, as this gas defuses much more slowly through water than through air. The isoetid-strategy plants have relatively small leaves but extensive root systems, and it is through these roots that they access carbon dioxide from within the sediments, where the levels of this gas can be raised by the respiration of microbes (Moss 2015). These plants grow slowly, but it's a strategy that works well in lakes where carbon dioxide is in short supply in the water body and where nutrients are also scarce. Many of our mountain lakes support such isoetid species, including the shallows of Windermere and other Lake District tarns, and many Snowdonian lakes including those around Cwm Idwal (Fig. 68). However, there is an alternative strategy used by other aquatic plants, the bicarbonate strategy.

Unlike the isoetid plants, which require a supply of carbon dioxide, another group of submerged macrophytes uses the bicarbonate strategy. This means that they get the carbon they need from bicarbonate in the water, which they can break down to release carbon dioxide. Lakes in catchments with more lime-rich rocks tend to have much more bicarbonate in their waters and are dominated by

FIG 68. Water Lobelia *Lobelia dortmanna* growing in around 1 m of water in Llyn Clyn, near Cwm Idwal in Snowdonia. Note the growths of epiphytic algae on the flowering stems growing up through the water body.

plants using this strategy (Fig. 69). The use of the bicarbonate gives them access to more carbon for growth, so they can outcompete plants that are solely using carbon dioxide from the water, especially in more nutrient-rich waters. Many of the macrophytes using the isoetid strategy have become rarer as we have tended

FIG 69. Malham Tarn in the Yorkshire Dales is the largest natural marl lake in Britain. Marl lakes are bicarbonate-rich water bodies where calcium carbonate is deposited on the lake bottom, in part due to chemical changes associated with photosynthesis by plants using the bicarbonate strategy.

to make lakes more nutrient-rich via pollution (the naiad in Esthwaite Water being one example). But over recent decades British lakes have seen an overall decline in macrophytes using both strategies, and in the next section we will see how these declines illustrate some of the complexities of ecosystem functioning.

This brief discussion of photosynthesis by submerged macrophytes helps explain why lakes can't be just treated as distinct microcosms, isolated from the surrounding world. The nature of the rocks in the catchment is clearly important to the relative success of isoetid versus bicarbonate photosynthetic strategies, as is the respiration by microorganisms in the sediments. In Windermere and nearby lakes pollution associated with the rise in tourism provides another example. In addition, things happening outside the catchment also affect aquatic photosynthesis. For example, carbon dioxide dissolves into the water from the air – as well as coming from respiration of organisms living in the lake – and humans are slowly increasing the atmospheric levels of this gas. As is well known, these changes are also affecting the climate, and warmer water contains less carbon dioxide (and oxygen too), so events far away from the catchment can still affect the lake. Clearly the abiotic environment is important: one cannot understand the natural history of these lakes – or of any ecosystem – if only the organisms are considered, and the physics and chemistry of their environment is ignored. The holistic 'from physics to fish' approach is more than a sound bite; it makes sense.

THE COMPLEXITIES OF ECOSYSTEM FUNCTION – MACROPHYTES, ALGAE AND ALTERNATIVE STABLE STATES

Moving away from the mountains, classic examples of shallow lakes with similar pollution issues to Windermere are the Norfolk Broads and the Shropshire and Cheshire meres (Figs. 70 & 71). The Broads have been particularly well studied by Brian Moss and his colleagues – and indeed much of the discussion in the next couple of paragraphs is influenced by these studies (particularly Phillips *et al.* 2016). There were large changes in the aquatic plants in the Broads following the Second World War, and in explanation it was suggested that increased boat traffic was stirring up sediment so the water was too cloudy to allow the light to reach the plants. More broadly, both in the Broads and in the other 'lake districts', another idea was that increased nutrients from pollution caused an increase in algal growth in the water body, also reducing light (Moss 2001). There is an element of truth in this idea, as shown by the changes in Secchi depth in Windermere and Esthwaite Water described above. However, things are more

FIG 70. Hickling Broad is the largest of the Norfolk Broads, a series of shallow lakes formed by the flooding of pits created by medieval peat digging during the ninth to thirteenth centuries. Hickling Broad is an Environmental Change Network site, and over the past 50 years the water clarity and extent of macrophyte coverage here has varied. The Broad has suffered eutrophication in the past, with the 1970s being a time of particularly high nutrient levels and low macrophyte cover, in part because of a gull roost which declined during the 1980s and 1990s. Conditions were good around 2008/09 but have since deteriorated somewhat with increasing nutrient levels (Moss 2001, 2015).

complex than this simple story would suggest, as there are many other reasons for turbid water besides algae, such as dissolved organic material and suspended inorganic matter (Fig. 72). Certainly most lakes tend to become dominated by phytoplankton as nutrient levels increase, but it seems likely that this is caused by the loss of macrophytes, rather than being the cause of plant decline.

How might macrophytes make it less likely that a lake becomes dominated by phytoplankton? One of the important mechanisms appears to be the classic idea of a food chain. Key predators of algae are many members of the zooplankton – such as the water fleas (Cladocera) and the copepods. These zooplankton are themselves food for predatory fish, and macrophytes provide cover in which zooplankton can hide from their predators. So as the cover of macrophytes decreases the populations of zooplankton can fall, with the result that predation on the phytoplankton decreases and the water becomes turbid and green from

FIG 71. Rostherne Mere is one of the largest of the Cheshire and Shropshire meres. Most of these meres are naturally nutrient-rich (eutrophic), but many, including Rostherne, have become more eutrophic over the past 100 years from agricultural changes, and in the case of Rostherne pollution from sewage treatment works too. Rostherne has a long history of ecological studies, with data on the freshwater algae going back to work by W. H. Pearsall just prior to the First World War. The mere became a National Nature Reserve in 1961 (Fisher *et al.* 2009, Wall & Wall 2019).

FIG 72. Phytoplankton are not the only reason for reduced transparency in water. The photograph shows the bed of the River Tees just upstream from Cow Green Reservoir (see the opening of Chapter 4 for a description of the area). In this case the reduced visibility is due to peat-stained water.

the increased algal populations. In addition, as nutrients increase, epiphytic algae expand on the surface of macrophytes, reducing the amount of light available to them for photosynthesis (Fig. 68). Several studies have suggested that the presence of grazing invertebrates (such as snails) can also be important, and that if there are good populations of grazers then such epiphytic algae are less of a problem to the macrophytes. So it seems that while higher levels of nutrients (mainly phosphorus and/or nitrogen) increase the likelihood of a switch to a lake dominated by algae, this often isn't a foregone conclusion, and that two alternative states are possible – namely a lake dominated by macrophytes or one dominated by plankton (algae) (Fig. 73). Other stresses on the system, such as other sources of pollution killing fish and/or grazing invertebrates, or uprooting of macrophytes by wildfowl (Fig. 74), may sometimes trigger a switch between these two states too.

FIG 73. Alternative stable states in shallow lakes. The panel below summarises mechanisms that help maintain submerged plant communities, while the panel on page 109 illustrates the mechanisms that help stabilise phytoplankton-dominated communities. (Redrawn from Moss 2001)

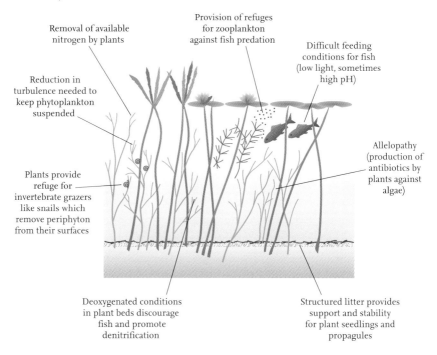

Removal of available nitrogen by plants

Provision of refuges for zooplankton against fish predation

Difficult feeding conditions for fish (low light, sometimes high pH)

Reduction in turbulence needed to keep phytoplankton suspended

Allelopathy (production of antibiotics by plants against algae)

Plants provide refuge for invertebrate grazers like snails which remove periphyton from their surfaces

Deoxygenated conditions in plant beds discourage fish and promote denitrification

Structured litter provides support and stability for plant seedlings and propagules

These ideas of alternative stable states (macrophytes or algae) have wider importance. The theoretical possibility of ecosystems switching between different states goes back to at least the 1970s, but lakes provided real examples of this influential idea and made it an active area of research (Scheffer *et al.* 1993). The basic idea can seem counterintuitive, and so it's easy to overlook the potential for such changes. Most people tend to assume that if you make gradual changes to a system then it will respond in a gradual way too – that is, gradually changing causes (such as a slow increase in nutrients) lead to gradually changing effects (such as a steady increase in the number of algae in the water). A classic example that is likely to be familiar to many people as a distant memory of school science experiments is the demonstration of Hooke's law. If you hang weights on the end of a metal rod (in practice a spring is often used, as the changes are more easily seen) then it will lengthen. There is a linear relationship, so as you increase the strain on the spring it gets proportionally

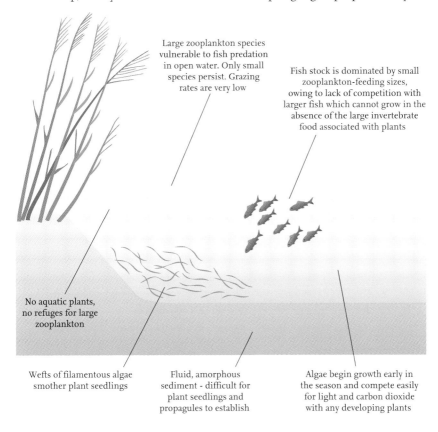

Large zooplankton species vulnerable to fish predation in open water. Only small species persist. Grazing rates are very low

Fish stock is dominated by small zooplankton-feeding sizes, owing to lack of competition with larger fish which cannot grow in the absence of the large invertebrate food associated with plants

No aquatic plants, no refuges for large zooplankton

Wefts of filamentous algae smother plant seedlings

Fluid, amorphous sediment - difficult for plant seedlings and propagules to establish

Algae begin growth early in the season and compete easily for light and carbon dioxide with any developing plants

FIG 74. Mute swans with their long necks can reach up to 1 m below the water surface and their grazing can have considerable effects on aquatic macrophytes in shallow waters (Wood *et al.* 2014).

stretched – draw a graph of length against weight and you should get a straight line. This is after all how a spring balance works. Equipment like this was commonly used to weigh birds during ringing before field electronic balances became readily available, and it is often still used for birds the size of larger waders and upwards (Fig. 75).

The working of a spring balance nicely matches the expectation that a gradually changing cause (the added weight) results in a gradually changing effect (the increased length). However, even in this simple experiment things become more complex – although your school science teacher may have stopped you before the results became more interesting to prevent you breaking the spring! At some point – technically called the 'proportionality limit' – the linear relationship starts to break down, although if you then remove the weights the spring (or rod) will return to its initial starting length. However, add more weight and you cross the 'elastic limit' – this has cost your school's science lab money, as the spring will no longer return to its original length and they will need to buy a new one. At a heavy enough weight – the aptly named 'rupture point' – you have not only straightened out the spring into a straight wire, you

FIG 75. Hooke's law in action: a spring balance being used to weigh a Whooper Swan *Cygnus cygnus* at the WWT's reserve at Martin Mere, Lancashire.

have snapped the wire too (Thorne & Blandford 2017). So even in this simple experiment, once you exceed the proportionality limit the nice straight-line relationship starts to break down, and if you exceed the rupture point then you suddenly jump to a very different system – a broken spring. Something similar seems to happen with increased nutrient concentrations in lake systems, as they jump to new stable states.

What about changing the system back from phytoplankton-dominated to one with clear water and macrophytes? This is a big issue for the management of many of the Norfolk Broads and similar lakes around Britain. Crucially, just reducing the amount of nutrients entering the lake to below the level at which the switch to phytoplankton dominance occurred seldom seems to be enough – the new algae-dominated system is stable (hence the term 'alternative *stable states*') and only substantially lower nutrient levels are likely to facilitate such a switch. One way to try and switch a lake back to a clear-water state is to modify other aspects of its ecology at the same time as trying to reduce nutrient levels. For example, some success has been achieved in parts of the Norfolk Broads by reducing fish numbers so that the populations of algae-eating zooplankton can increase (Moss 2001).

An additional problem, in both the Broads and elsewhere, is that lake sediments can form a reservoir of high nutrients which can re-enter the water.

So that even if nutrient levels entering the lake are reduced this will not necessarily lead to lower levels in the water. A good example of this problem can be seen at Rostherne Mere in Cheshire (Fig. 71). The Mere received substantial amounts of nutrients between the 1930s and 1991 from a sewage works which discharged into a stream which fed into the lake, along with another that discharged into the catchment further upstream. This led to high phosphorus levels in the sediments, which are likely to have substantial effects on the water chemistry for the foreseeable future (Radbourne *et al.* 2019). In addition, the Mere has other sources of nutrients entering it from surrounding agriculture, and from waterbirds (Fig. 76). The Mere is a well-known birdwatching site, and in the 1960s it was thought that the large number of waterbirds might have been the major source of nutrients into the lake. However, later work by Brian Moss and others showed that the birds had a very minor effect, with most nutrients coming from sewage effluent and agriculture (Wall & Wall 2019).

Recently this idea of alternative stable states and non-linear changes in ecosystems has become more widely applied under the title of 'tipping points', especially in relation to global climate (Lenton 2016). The suggestion is that the sort of rapid transitions that can happen in lakes – between macrophyte-dominated and phytoplankton-dominated – can also happen at much larger scales, and can be similarly difficult to reverse once they have occurred. Some of these tipping points are mainly based on physics rather than living organisms – for example, the loss of sea ice in the Arctic. In this case white ice reflects a lot of heat back to space, and as ice becomes less extensive because of increasing temperatures, less incoming heat is reflected and so temperature increases further – potentially leading to the loss of all sea ice. A more ecological example is provided by the forests of the Amazon. Much of this forest is a long way from the sea, and one might think that it would be short of rain as it's too far from the ocean to be reached by most clouds – which are likely to have already shed their rain nearer the coast. However, large amounts of water evaporate from forest (both from rain that has fallen on leaves and then evaporates, and as transpiration – water that has been taken up through the roots and is eventually lost from the leaves). So the forest is continually returning water to the atmosphere, which can then fall again as rain. If enough forest is cleared and/or the climate changes enough, then potentially a tipping point is reached when not enough water is being returned to the atmosphere from the vegetation to supply the necessary rainfall, and the forest can suddenly become unsustainable.

FIG 76. Input of nutrients from Cormorants *Phalacrocorax carbo* in a lakeside woodland at Rostherne Mere. The Cormorants both roost and breed in the trees above, making the sedges white with bird faeces. When I first knew this site as a teenage birdwatcher in the late 1970s there was a winter Cormorant roost in these trees with only the occasional bird in summer and no breeding colony. The birds started breeding here in 2004, increasing to 168 nests by 2014 (Wall & Wall 2019). The photograph was taken in June 2016, when I counted at least 220 birds in the trees above – a spectacular change from when I first knew the site.

NUTRIENT CYCLING

Much of this chapter so far has used photosynthesis in lakes, along with human-caused increases in some nutrients, as a way of starting to think about how ecosystems might work. One way of thinking about the ecology of lakes is to consider a lake as analogous to an organism – which can be either autotrophic or heterotrophic. An autotrophic lake would be one where enough energy is captured by photosynthesis within the lake (by macrophytes and phytoplankton) to support all the life in the lake, whereas a heterotrophic lake is one where the life in the lake is utilising more energy than can by fixed by within-lake photosynthesis, and it is subsidised by energy coming in from the catchment. An example of this would be subsidies from organic matter generated by photosynthesis by terrestrial plants in the lake's catchment. Traditionally lakes were often thought of as autotrophic (e.g. the idea of a lake as an island microcosm), but it is now realised that many are more likely to be heterotrophic (Moss 2010). Indeed, to be accurate we should really think of these lakes as analogous to mixotrophs (organisms which are both heterotrophic and photosynthetic – see Chapter 7), since the energy is coming from a mix of within-lake photosynthesis and the wider catchment.

Moving from thinking about energy to considering nutrients, the idea of cycling becomes important. Living organisms require a wide range of chemical elements to function – 20 or so are needed for life, with carbon, hydrogen, nitrogen, oxygen, phosphorus and sulphur (the oft-used mnemonic is CHNOPS) usually the ones needed in largest quantities. The two key nutrients in lake eutrophication are nitrogen and phosphorus. However, as well as the CHNOPS group other elements can be important too. For example, in many of the lakes discussed in this chapter microscopic diatoms (Fig. 59) require silicon in some quantity to make their cases, and this can sometimes be in short supply, while iron can be limiting for plankton growth in many marine systems away from the coasts.

An idealised lake, with inflowing streams and a river as outflow, makes the potential importance of nutrient cycling clearer than is the case in an ecosystem such as a woodland. If there is no nutrient cycling, then the maximum amount of life that can live in the lake will presumably be limited by the inflowing nutrients. If nutrients either flow through the lake and are lost in the outflow or are used just once by an organism before being lost to the outflow or lake sediments, this puts a limit on the amount of life that the lake can support. However, if nutrients can be used multiple times by life before being lost from the lake, then potentially much more life can be supported. This applies to any ecosystem, but the point is more easily appreciated for lakes because of their apparently obvious boundary.

A good way of understanding the importance of nutrient cycling is to look at the concept of the cycling ratio (Volk 1998). This is the ratio between the amount of a nutrient passing through life within a system (such as a lake) and the amount which exits the cycle. For example, in terrestrial systems the cycling ratio of phosphorus is around 46 – that is, for every 46 atoms of phosphorus currently active within the cycle around one is lost from the system. So instead of each atom being used by life just once before being lost, on average it is used 46 times. In marine systems the cycling ratio for phosphorus is even higher, with a value of around 280. So nutrient cycling allows the support of far more life than if cycling didn't happen in a system. In computer models of life, which allow the virtual organisms to evolve to use different virtual nutrients, nutrient cycling readily develops as organisms evolve to utilise waste products produced by other organisms (Downing & Zvirinsky 1999, Williams & Lenton 2007). Indeed, because of this I have previously argued that nutrient cycles are likely to be a fundamental part of the ecology of any planet with life (Wilkinson 2006).

If we switch back from these broad theoretical generalisations to considering the natural history of a particular location, then Rostherne Mere in Cheshire illustrates a number of interesting complications (the following is mainly based on Reynolds 2019). The lake is naturally nutrient-rich and has become even more nutrient-rich because of pollution. So it provides a nice example of a lake where nutrients do not seem to be limiting the amount of life – this is because in the green, algae-rich water light for photosynthesis becomes limiting before either nitrogen or phosphorus is in short supply. At times during the summer the Secchi depth can be as little as 1–2 m. Another complication is that in the case of nitrogen some of the phytoplankton can be a source of the nutrient too. Early in the year in Rostherne there is usually a spring bloom in diatoms, followed later in the summer by large numbers of either dinoflagellates or colonial cyanobacteria. These are often followed in late summer by large numbers of filamentous cyanobacteria, such as *Anabaena*. Similar patterns are found in many other lakes, including the well-studied lakes of southern Cumbria.

The cyanobacteria are particularly interesting because of the ability of some species to fix nitrogen. Nitrogen is crucial for life – for example, it is used in making protein – but it's hard for organisms to access. This appears strange, given that nitrogen is the commonest gas in Earth's atmosphere, but most of the nitrogen in the atmosphere is in the form of dinitrogen (N_2) molecules, which are very stable, and only a few organisms can access nitrogen in this form. Some cyanobacteria are amongst the small number of organisms that can do this. With the exception of human chemists, all nitrogen fixers are prokaryotes.

Cyanobacteria (in the past often called blue-green algae) are photosynthetic prokaryotes which are found in a wide range of environments. Many species can fix nitrogen – that is, access the dinitrogen molecules in the atmosphere and turn them into nitrogen compounds which are accessible for use by themselves and a wide range of other organisms. This leakage of nitrogen into the environment from a nitrogen-fixing organism is not just an aquatic phenomenon, and it is described in a terrestrial context for cyanobacteria in dog lichens in Chapter 7.

There is a serious difficulty for photosynthetic organisms like cyanobacteria when fixing nitrogen. Key chemicals (nitrogenase enzymes) used in the fixation process are damaged by oxygen, but oxygen is given off as a waste product of photosynthesis. The way many freshwater cyanobacteria get round this is to conduct nitrogen fixation in specialised cells called heterocysts which keep oxygen away from their nitrogenase (Fig. 77); however, many marine cyanobacteria sidestep the oxygen problem in ways that don't require heterocysts, which has until recently caused us to underestimate the amount of nitrogen fixation in the ocean. The idea of using a specialised cell such as a heterocyst is counterintuitive if you are used to thinking of bacteria as single-celled organisms. The heterocyst-using cyanobacteria are *multicellular* organisms composed of chains of cells. However, there is also a cost to nitrogen fixation, as it requires a lot of energy (Whitton & Potts 2012), so it is usually only advantageous when biologically accessible nitrogen is in short supply in the environment.

Freshwater cyanobacteria tend to get most attention when their populations expand to very high numbers, so-called algal blooms. Blooms are usually dominated by only one or two species. In fact, 'bloom' itself is a poorly defined term just referring to a phytoplankton biomass significantly higher than average for a particular lake, often turning the waters a blue-green colour – although for drinking-water reservoirs there is often a somewhat arbitrary value for the amount of phytoplankton in a cubic metre above which a bloom is considered present (Oliver & Ganf 2000). Many cyanobacteria associated with these blooms release harmful toxins (Fig. 78). In the summer of 1978 three cows out of a herd of 40 died after drinking from Rostherne Mere during a bloom, probably the first recorded case of cattle death in Britain caused by cyanobacteria (Reynolds 2019). Such blooms are clearly an important issue when it comes to more applied aspects of freshwater ecology; for example, their effects have been estimated to cost $4 billion annually in the USA alone (Ho *et al.* 2019). However, the main focus of this book is on the underlying ideas of ecology, and blooms are instructive here too.

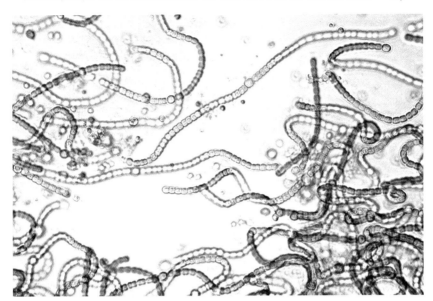

FIG 77. Cyanobacteria (*Anabaena* sp.). The larger-diameter cells in the centre of some of these chains are the heterocysts (sites of nitrogen fixation), with a diameter of around 7.5 μm. These cyanobacteria were cultured from Mute Swan faeces collected from Windermere (see Fig. 62) and have most likely developed from cells that had passed through the swan in a living state (although it was not possible to rule out the slim chance that airborne cyanobacteria might have contaminated the culture).

FIG 78. Warning sign for a summer cyanobacterial bloom at Malham Tarn in the Yorkshire Dales (see Fig. 69).

Globally, cyanobacterial blooms are becoming more common. In a study of 71 very large lakes around the world (the size meant there were no UK examples in the study) Ho *et al.* (2019) found that bloom intensity had increased since the 1980s in 68 per cent of the lakes studied. They found few obvious patterns in these data that explained this rise, except that the lakes with reduced bloom intensity had warmed less than other lakes. Indeed, increased temperature is thought to be one of the two main factors driving algal blooms, the other being increased nutrients. Obviously these two factors interact, and it has been suggested that if you want to counteract the effects of increasing temperature on blooms in a lake or reservoir then you need to reduce nutrients by one-third to offset a 1 °C warming, although this is obviously a very ballpark figure (Scheffer *et al.* 2015).

The interactions between algal blooms, temperature and nutrients provide a nice illustration of the fact that in ecosystem ecology you often have multiple interacting factors affecting the system you are studying. Back in 1999 John Lawton argued that some of the hardest aspects of ecology to understand were those at these intermediate scales, with population ecology (Chapter 8) having less complexity and so being easier to grasp – although he did think that lake systems were probably less complex than the terrestrial ones he tended to study (Lawton 1999a). He also thought that at larger more global scales things might be more readily understood, as some of the detail and the complexity is lost when you look at very large patterns. In the next section I will increase the scale and briefly consider the global ecosystem, before returning to use some of the ideas to examine a number of issues concerning the stability of the smaller ecosystems ecologists usually study.

THE PLANETARY ECOSYSTEM – A GAIAN VIEW OF ECOLOGY

In his 1887 'lake as a microcosm' paper, Forbes suggested that lakes were relatively stable systems because natural selection had acted on the inhabitants of a lake to allow them to evolve a system in a 'harmonious balance' where no species became too dominant. However, as I pointed out earlier in this chapter, from the 1960s onwards biologists became increasingly sure that such explanations can, at best, only work in rather special circumstances, and that natural selection by itself struggles to explain the evolution of benefits to groups of organisms of different species. This raises a series of questions. Why do ecosystems appear relatively stable? Is this apparent stability illusory? Is it due to chance, or are there mechanisms that promote stability in such systems?

The short answer is that these questions are very much at the cutting edge of ecology and so there is as yet no widely agreed answer. The whole surface of the Earth is a rather special system, as almost no matter enters it from outside, or indeed leaves it – the technical term is that it is 'materially closed'. This obviously contrasts with many of the Earth's smaller component ecosystems, such as lakes with their inflow and outflow of organic matter and nutrients. In a system that is almost 100 per cent closed, nutrient recycling is obviously extremely important. However, although it is closed to matter, it is open to energy. Like the smaller-scale ecosystems, the planetary one has large amounts of energy (light) arriving from outside which is utilised by photosynthetic organisms.

What should we call this planetary ecosystem? In general I don't get particularly excited about questions of terminology, but because of the focus on planetary-scale problems – such as climate change – in this case it's a question of some interest. Richard Huggett (1999) suggested that the terminology was somewhat confused but that three possible terms were 'biosphere', 'ecosphere' or 'Gaia'. The one most commonly used in ecology texts is biosphere – but there is a lack of consistency in how the term is defined. In some cases 'biosphere' refers to the totality of life on Earth, while in other cases it means the whole system of life and the life-support systems of atmosphere, water, rocks and soils. In effect it is this second 'systems' definition of biosphere that describes the global ecosystem, while the other definition, with its focus on life, is defining the biosphere as a global ecological community. To avoid this confusion Huggett preferred the term 'ecosphere' for the global ecosystem; however, 'biosphere' is the more familiar term, and from early in its usage, in the early twentieth century, it was employed by the Russian scientist Vladimir Vernadsky in a systems sense – and it seems reasonable to continue with this usage. Much of the remainder of this chapter discusses the Gaia hypothesis without addressing the issue of the terminology. I will return to the term 'Gaia', and the extent to which it is sensible to use it as a synonym for the systems definition of biosphere, in this chapter's concluding remarks.

One of the most striking facts in ecology is the long geological history of the biosphere. The date of the first life on Earth is uncertain, for fairly obvious reasons – we are looking for something very small and simple in the fossil record from a time represented by very few surviving rocks. However, for the last couple of decades there has been a reasonable consensus that although the exact timing is still unknown, Earth has been a living planet for at least 3.5 billion years, probably closer to 4 billion, and possibly even a bit longer than that (Schopf 2006, Lenton & Watson 2011). This long history of life on our planet raises a fascinating ecological question, namely why has it survived so long despite a myriad things that could have made the planet uninhabitable? For example, think of the narrow

TABLE 2. Maximum and minimum temperatures at which different groups of organisms can complete their life cycle (data from Clarke 2017). Organisms can often survive more extreme temperatures than these in a dormant state: for example, some bacteria have been described surviving temperatures at least as low as –196 °C. The high maximum value of 60 °C for animals is for nematode worms in compost heaps. In general, over time, these data tend to change, becoming more extreme as we find new examples. For instance, several of the maximum values in this table are higher than the ones in a similar table I compiled in the mid-2000s (Wilkinson 2006). However, life on Earth in general, and larger multicellular organisms such as plants and animals in particular, appears to be able to live only over a relatively narrow range of temperatures. One way of phrasing the Gaia question is to ask, is it chance or some more interesting mechanism(s) that has maintained life-friendly conditions on the planet for around 4 billion years?

Group	Max temperature (°C)	Min temperature (°C)
Archaea	122	–16.5
Bacteria	100	–20
Single-celled eukaryotes	60	–8
Yeasts	60–62	–18
Lichens	45?	–10
Animals	60	–2
Vascular plants	65	0

range of temperature over which life can survive (Table 2). In nearly 4 billion years, there seems a substantial chance of the Earth becoming either too cold, or too hot, for life. There are several potential answers to this question, including that it is entirely down to luck that life has survived so long, or alternatively that there could be something about planets with abundant life that makes them more likely to maintain life-friendly conditions. This second possibility is one way of stating the Gaia hypothesis, first suggested by James Lovelock (Fig. 79) in the 1960s and 1970s and developed with important early microbiological and geological inputs from Lynn Margulis (Fig. 53).

In a modern formulation, Gaia suggests that a coupled system of life on Earth and its abiotic environment self-regulates in a habitable state, despite destabilising influences such as a steadily brightening sun, changes in geological activity, and occasional massive meteorite impacts (Lenton *et al.* 2018). There are two key technical ideas in this formulation. First, a 'coupled system' is the idea discussed in Chapter 2 that life affects the abiotic environment, as well as the abiotic environment affecting life. Therefore the two need to be considered

FIG 79. James Lovelock, the originator of the Gaia hypothesis, photographed in 2011. For an idea of how his ideas developed over time compare Margulis & Lovelock (1974) with Lovelock (1995, 2003). For more critical (i.e. less convinced of the idea) discussions of Gaia see Tyrrell (2013) and/or Waltham (2014) – although note that both of these were written before the publication of recent theoretical work by Doolittle and others described in the text.

together as they affect each other – in the context of the main examples discussed in the current chapter it's an extension of the idea of a lake as a holistic system 'from physics to fish'. The second idea is 'regulation' (described in more detail in the context of population ecology in Chapter 8), meaning that the system has some thermostat-like qualities tending to keep it relatively stable in the face of external changes. This leads to a certain amount of stability in conditions on Earth – although the extent of this stability is considered uncertain even by most of those scientists who think Gaia a reasonable way of thinking about ecology at a planetary scale. One difficulty is the question, what exactly would you expect from a regulated system? There seems to be an obvious answer – if you have regulation then things don't change over time. However, a moment's thought shows that things are a bit more complex than this simple answer suggests. Think of a hot shower. Ideally what you want from a shower's thermostat is that it keeps the temperature of the water at whatever level you set it. However, stand under an old shower and you may experience the water getting hotter, then colder, then hotter again. It's still regulated around an average temperature but it's hardly constant.

In thinking about Gaia it's important to appreciate that it is not really a single scientific theory but a collection of scientific ideas, which mesh together to provide a framework for understanding the world. In a similar way Darwin's ideas on evolution are also often described as a set of theories – such as the idea of evolution as such, common descent of all life from a single ancestor, the mechanism of natural selection, and more (Mayr 1991). Some of the aspects of Gaia that were considered controversial in the 1970s are now mainstream – for example, the close coupling of life and the abiotic environment. As Lovelock pointed out in many of his publications, the Earth's atmosphere would have a very different chemical composition if there was no life on the planet, for example containing almost no oxygen. This now forms part of what's called Earth Systems Science, and as the ecologist John Lawton pointed out in 2001:

> *James Lovelock's penetrating insights that a planet with abundant life will have an atmosphere shifted into extreme thermodynamic disequilibrium, and that Earth is habitable because of complex linkages and feedbacks between the atmosphere, oceans, land and biosphere, were major stepping-stones in the emergence of this new science.*

However, the idea that this coupling leads to a thermostat-like regulation has remained more controversial. The obvious question that occurred to many biologists is to ask how such a thing could evolve. Early versions of Gaia appear to assume that natural selection will lead to organisms doing things that are to the benefit of life as a whole, for example plants producing oxygen for the common good. The problem is that one would expect an organism that ignored this role and just followed its self-interest to do better than those spending some of their resources for the benefit of all, so the system would break down as the self-interested outcompeted the selfless. Many people made similar points about Gaia around the start of the 1980s, with arguably the two most influential being commentaries by Ford Doolittle (1981) and Richard Dawkins (1982). Part of the answer lies in the example of oxygen production by plants 'for the good of the planet'; indeed, this was one of the examples used by Dawkins in his critique of Gaia. However, oxygen is a by-product of photosynthesis rather than being produced directly for its Gaian effect. Towards the end of the 1990s several people pointed out that if Gaia is built of by-products this removes some of the evolutionary problems – although it was not obvious why these by-products should have stabilising, life-friendly, functions (e.g. Volk 1998, Wilkinson 1999a).

In the last few years a number of ideas have been developed that may help explain the long-term stability of life on Earth, and potentially of smaller ecosystems too. I briefly introduce these ideas next, but it should be remembered that many of them are new and designed to address really complex and difficult questions. It's not yet obvious how useful they will turn out to be, and in addition, since I have been involved in some of this work it's quite likely I am somewhat biased in favour of these ideas. One way of thinking about this is to ask the question 'How would the following paragraphs be rewritten in a hypothetical second edition of this book written in 10–20 years' time?' One possibility is that these ideas won't feature in the rewrite at all because ongoing research has suggested they don't work, or alternatively they will be presented simply as textbook orthodoxy. Most likely the position will be part way between these two extremes, with at least some of the ideas considered partial solutions to explaining both Gaia and the workings of smaller ecosystems.

There is now a range of possible mechanisms that may explain how Gaia works (Lenton *et al.* 2018). As is often the case in ecology it's unlikely that just one of these will turn out to be 'The Answer', with the others all being wrong. More likely is that several of these ideas contribute to the correct explanation, with the question then becoming which are the most important, and which just make minor contributions to the answer. Conventional natural selection can contribute, although it can't provide the total answer, for the reasons outlined above. For example, it's straightforward to visualise organisms evolving to make use of underutilised sources of energy and nutrients in the environment (such as waste products of other organisms, be they faeces or discarded leaves). From this, nutrient cycling emerges, helping to maintain a long-term source of nutrients for life. It's also possible to see how nitrogen fixation in lakes, or the ocean, could be regulated in line with conventional natural selection. The fixing of nitrogen is obviously advantageous to the cyanobacteria that can do this, and some leakage into the surrounding water is likely unavoidable – thus benefiting other non-fixing organisms. If this leakage is great enough, these other organisms do well and compete with the cyanobacteria, so reducing their populations and reducing the leakage of nitrogen compounds into the environment.

The really interesting ideas are, however, ones that don't neatly map on to conventional ideas about natural selection. For example, in the case of nutrient cycles there is the possibility that focusing the explanation on the evolution of the organisms themselves – the natural approach for most biologists – may be missing important aspects, and we should consider instead the evolution of the cycles themselves. In Ford Doolittle's memorable phrase, we should be thinking about 'the song, not the singers' (Doolittle 2017). Potentially the cycles can evolve, even if the organisms involved in them change over time, just as a song can stay the same although the people in the choir change over time. Doolittle (2014) has also suggested that what he calls 'selection through survival alone' may form part of the theoretical explanation for Gaia.

The fact that many of the recent ideas about how Gaia may work come from Ford Doolittle is interesting, given that, as pointed out above, a few decades ago he wrote one of the better-known essays explaining why Gaia couldn't work (Doolittle 1981). Ford has told me that the reason he returned to thinking about Gaia was that he became dissatisfied with recent suggestions on the evolution of the microbiome (the total genetic material of all microbes living in another organism – such as a human – a topic of increasing medical interest). He realised that many of the problems he had with these ideas were very similar to the problems he had with Gaia, and so started to think about alternative approaches.

One obvious problem with applying evolutionary ideas to Gaia (or indeed the microbiome) is that it doesn't obviously reproduce, creating multiple genetically related individuals upon which natural selection can act. The key idea of 'selection through survival alone' is that the longer a system survives the more chance it has of acquiring some particular stabilising mechanism – say perhaps something to do with nutrient cycling. The acquisition of this mechanism increases the system's chance of survival, as it is now more stable, so also increasing its chance of acquiring additional stabilising mechanisms, since it can survive for longer while waiting for improvements to arise. In Doolittle's original version of this idea the argument was largely verbal, but recently a first go at trying to turn it into a more mathematical form suggests that the idea looks plausible (Nicholson *et al.* 2018). An alternative way of a system acquiring stability is 'sequential selection', where the system can have multiple goes at finding a stable state – if the system isn't stable it quickly collapses and, unless this leads to the total extinction of life, it can try repeatedly until a more stable configuration is found. It's likely that fluctuations that fall short of total collapse may also play a role in acquiring stability (Lenton *et al.* 2018). By definition more stable systems last longer than unstable ones, however they have acquired their stability, so it's not too surprising that the systems we see in ecology look reasonably stable.

All these potential mechanisms for increasing stability are relevant to conventional ecosystems as well as the biosphere. In the early days of the study of ecology of lake systems, Stephen Forbes suggested that natural selection would lead to the evolution of a stable system for the benefit of all the organisms living in the lake, and early versions of Gaia often implied something similar. As described above, we are now confident that this is not usually the case, but we now have a number of emerging ideas which seem to offer more plausible mechanisms for stability. What we currently don't know is the size of any effect these mechanisms can have on stability. Are they so minor that they are of very limited importance, or does a major part of the apparent stability of life on Earth over geological time come from some combination of these (and other unknown?) mechanisms?

CONCLUDING REMARKS

Ecosystems are one of the fundamental things – or entities – in ecology, comprising not only all of the organisms in that system but also its physics and chemistry and all the interactions and feedbacks between these many components. This can be contrasted with the idea of an ecological community, which comprises just the organisms. In some cases focusing just on a community may be appropriate (for example, when discussing how insect communities differ between sites with different plant communities) – but for most questions ecologists are interested in it makes little sense to ignore the interactions with the wider environment. This chapter has introduced ecosystems mainly using the examples of lakes. There are good reasons for this. First, lakes have a more obvious edge than most systems, although this simplicity turns out to be somewhat illusory, as it's hard to understand the functioning of most lakes without considering their whole catchment. Second, many influential early studies that gave rise to ecosystem ecology used lakes, classic mid-twentieth-century work by Lindeman and Hutchinson being obvious examples. Following from this, a holistic systems approach has arguably been more common in freshwater ecology than in other areas of the subject – the so-called physics-to-fish approach. This systems (or Earth Systems) approach to ecology now looks central to our attempts to understand the scale and impact of humans on our planet.

The difficulty with studying ecosystems is that they are very complex, with all the many interactions between life and the abiotic environment. The examples of non-linear behaviour and alternative stable states in lakes discussed in this chapter provide clear examples of this difficulty. When I was a university student in the 1980s ecosystem ecology was experiencing reduced research interest, with far more work being focused on simpler systems more amenable to experimentation. In 1989 the freshwater ecologist Brian Moss was concerned that ecologists were tending to ignore big-picture ecosystem studies, leaving such questions to chemists and other environmental scientists – and although things have improved somewhat, there is still an element of truth to this observation today. Nonetheless, during the 1990s interest in ecosystem ecology started to increase again, partly driven by worries about the large-scale effects of humans on the global environment, and partly because new technology (fast computers, satellite remote sensing, and techniques imported from molecular biology) made it easier to address some of these questions (Wilkinson 2006, Grace 2019). For example, many of the more recent studies of aquatic microbes discussed in this chapter have made use of molecular techniques, and the global study of cyanobacterial blooms by Ho *et al.* (2019) was based on remote sensing data.

Similar points were made in Chapter 4 on biodiversity, namely the growing importance of molecular methods especially in microbial ecology. Many of these microbes are involved in recycling loops, and many ecosystems are remarkably effective at recycling nutrients multiple times before they are lost from the system. Tim Lenton and Bruno Latour (2018) have made this point succinctly: 'The participants in the recycling loop are no longer limited by what comes into their world, but rather by how efficiently they can recycle resources.' They go on to make the point that currently humans are much less good at recycling than the rest of nature. We have much to learn from ecosystem ecology.

At a very large scale the whole surface of the Earth makes a global ecosystem. This is usually termed the biosphere in ecology texts, and despite some confusion with multiple definitions of this term it's probably a good name for our global ecosystem. I have some sympathy with the idea of using Gaia as a synonym for the biosphere, as it has a nice poetic ring to it. However, Gaia is probably better reserved for a particular theoretical approach to understanding the long-term survival of the biosphere. Its key idea is that through the interactions between life and the abiotic environment a certain (currently hard to quantify) amount of stability is introduced to the global ecosystem. This suggests a view of ecology with a somewhat different emphasis from that traditionally taken by ecologists. For example, it implies a greater emphasis on processes rather than species, and a realisation that many of the key players are microbes, fungi and plants rather than the animals which often dominate ecology textbooks.

CHAPTER 6

Competition on the Isle of Cumbrae

It is mid-tide and I am sitting, tucked out of the wind, amongst lichen-splattered rocks looking out over the sea. I am nestled at the end of Farland Point on Great Cumbrae in the Firth of Clyde, with the mountains of Arran looking enticing in semi-silhouette off to my right. On the rocks around me is scattered the clutter of a natural historian – field guides, camera equipment, binoculars, notebook and hand lens. Out to sea to the south is the island of Ailsa Craig, the likely source of the Gannets I can see plunge-diving for fish in the distance. However, the animals that have drawn me here are far smaller – the barnacles on the rocks that are steadily being uncovered by the retreating tide. Just below me the rocks are painted black with *Hydropunctaria* lichens, while below this they start to be crowded white with barnacles, along with the occasional tuft of seaweed. In the still-retreating water, the tops of kelp fronds are starting to break the surface – it is a particularly low tide, and slowly the full range of intertidal algae is emerging. As I scramble down onto the rocks, it becomes clear that there are at least two types of barnacle here.

The barnacles of Farland Point are famous, as this is probably the best-known barnacle research site in ecology (Fig. 80). Classic studies carried out here in the 1950s by Joe Connell still feature in the chapters on competition in the majority of university ecology textbooks. This chapter uses these barnacles, along with other examples, to introduce some of the basic ideas of competition in ecology, ideas that will be returned to and expanded on in subsequent chapters.

Farland Point is just outside Millport, the largest human settlement on the island of Great Cumbrae. The beach in Millport is a mix of sand, rocks and rock

FIG 80. Farland Point, Great Cumbrae, looking west across Millport Bay. This was the site of Joseph Connell's classic barnacle experiments during the 1950s.

pools, a habitat that attracts naturalists of all ages, including many who wouldn't recognise themselves in such a description but are still fascinated by seashore life. More serious students come here too – a short walk from Farland Point is the Field Studies Council's Millport Centre, an important location in the history of British marine biology. The first marine biology laboratory on Cumbrae was a 25 m boat moored there in 1884 by Sir John Murray (this was one of the first dedicated marine biology labs in Britain). The original buildings of the Millport Marine Biological Station were built in 1896, and by the start of the First World War this had become the home of the Scottish Marine Biological Association – which remained based here until it relocated to Oban in 1970. Following this the labs were run by the University of London until taken over by the Field Studies Council in 2014 (Archer-Thomson & Cremona 2019). At Millport today there are new student accommodation blocks, a small aquarium open to the public, and multiple labs with sea water literally on tap for visiting students to use (Fig. 81). So when Joe Connell was carrying out his classic barnacle experiments here (from 1952 to 1955), Millport was an active marine field laboratory. However, before describing the details of Connell's work it will be useful to briefly introduce some key ideas, to give his work context.

FIG 81. The Field Centre at Millport. The buildings shown in the photograph were part of the original Marine Biological Station opened in 1896. The earliest building is the wing to the left, with the right-hand wing added later. There are now a range of newer buildings off to the right, many built after the Field Studies Council took over the site in 2014.

WAYS OF COMPETING

A typical definition of competition would be two individuals attempting to use the same limited resource, often to the detriment of them both as even an individual that wins has had to spend resources to be the victor. These competing individuals may be members of the same species (in which case it's referred to as intraspecific competition) or come from two different species (interspecific competition). It's usually assumed that competition is greatest between members of the same species, as these tend to have the same requirements – for example, food requirements in the case of animals, or soil chemistry in the case of plants. Because they stay rooted to the spot, plants make good starting examples to help clarify ideas without the confusion of movement. Consider the familiar example of a clump of Daisies growing in a lawn (Fig. 82). Amongst other things they require space to grow, light for photosynthesis and nutrients from the soil. So a more formal definition of competition in plants is 'the tendency of neighbouring plants to utilise the same quantum of light, ion of mineral nutrient, molecule of water, or volume of space' (Grime 2001). The key point about this definition of Phil Grime's is that an individual quantum of light or molecule of water, etc., can only be used by one plant; hence the potential for competition for access to

FIG 82. Daisies *Bellis perennis* growing in a lawn. Competition can be easier to visualise in plants, as it is obvious that they compete for space, as well as for access to light, while their roots are potentially competing for water and nutrients.

the resource. The suggestion of Grime and his colleagues is that some plants that grow in productive, undisturbed habitats are adapted to be good competitors, growing quickly to try and monopolise these plentiful resources. Some are so successful they often form large single-species stands, such as reedbeds or nettle patches (Fig. 83).

Competition is often further subdivided into two main types, namely scramble competition and interference competition. These terms can apply to both intraspecific and interspecific competition. Scramble competition (also called resource, exploitation or explorative competition) is where a number of individuals use a common resource that is in short supply – such as the gulls fighting over the remains of someone's lunch in Figure 84. There are no all-out winners or losers, but all are ending up with less food than one gull would get if it was feeding alone at the café table. With interference competition (also called contest competition), some individuals do win out and acquire the resource at the expense of others, who get nothing (Krebs 2009). Examples of this sort of competition include the lichens and mosses competing for space on a rock in

FIG 83. Two examples of highly competitive plant species that often form single-species stands – both classified as 'competitors' by Grime *et al.* (2007). Top: Common Reed *Phragmites australis* at Cley Marshes in Norfolk. Bottom: Common Nettle *Urtica dioica* in a Lincolnshire woodland.

FIG 84. Scramble competition: Herring Gulls *Larus argentatus* raid the remains of someone's lunch on a café table. No single gull is managing to monopolise all the food, but all are likely ending up with less food than would be the case without competition from other gulls.

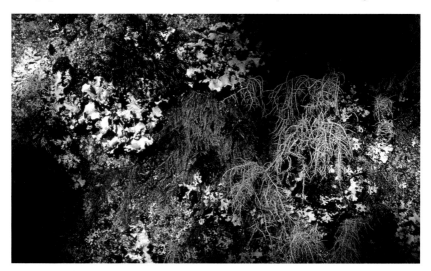

FIG 85. Interference competition in action in Wistmans Wood, Dartmoor. Lichens and mosses competing for space on a rock surface. The prominent leafy lichen is *Hypotrachyna laevigata*, while the branch-like lichens to the right are *Usnea* sp.

Wistmans Wood on Dartmoor (Fig. 85), or barnacles competing for limited space on the rock surface on a rocky shore, as described below.

Not all resources are subject to competition; oxygen is so plentiful in the air that organisms don't usually compete for it, since there is more than enough to go round. However, under some conditions in aquatic systems oxygen can become limiting, and then competition for it may happen. In addition, although many examples of competition have the organisms competing directly at the same time and place, it's possible for organisms to compete even if they never meet. Think of herbivores using the same food plant but feeding at different times of the day, or the examples of sperm competition described later in this chapter, where sperm from different males who copulated at different times compete to fertilise an egg.

In addition to these various types of competition, ecologists also sometimes write about 'apparent competition'. This is where two (or more) species have a disease or predator in common. If this causes more mortality to one of the species than the other it can create a distinct advantage for the less susceptible species. A well-studied example of apparent competition is between squirrel species in Britain, a topic discussed later in this chapter. Having introduced these various key ideas, we can return to look at the rocky shore at Millport and consider Connell's mid-twentieth-century barnacle studies.

COMPETITION AMONG BARNACLES ON CUMBRAE

Joe Connell was an American scientist who, after serving in the military during the Second World War, did a degree in meteorology before moving into biology. As a Masters student he worked on the American Brush Rabbit *Sylvilagus bachmani*, then after a spell as a biology teacher he moved to Scotland to do an ecology PhD based at the University of Glasgow (for accounts of the historical background for Connell's work I have mainly relied on Connell 2002 and Grodwohl *et al.* 2018). His experience with Brush Rabbit research led him to consider vertebrates in the wild far too difficult to work with, as they were hard to catch and it was very hard to amass a decent-sized data set. Because of this, and inspired by accounts of field experiments carried out at the start of the 1930s by Harry Hatton in France, Connell decided to work on barnacles. They have many advantages: the adults stay put, stuck to the rocks, and because they are small it's relatively easy to collect large amounts of data from a small area. In this sense they are more like plants than animals when it comes to the ease of data collection. The other advantage was that there was already a long history of barnacle research at Millport, providing lots of pre-existing data to build on.

If you examine the barnacles on a rocky shore at low tide with the aid of hand lens and field guide, you will usually find there is more than one species present, and that different species tend to be found at different heights on the rocks. This is certainly the case at Farland Point, where star barnacles tend to be commoner towards high tide levels while acorn barnacles are commoner lower on the shore (Fig. 86; Box 1). An interesting question is what governs where you find the different barnacle species – for since their larvae are planktonic you might expect them all to have an equal chance of colonising any part of the shore. Certainly it's unlikely that they could deliberately select where to colonise, as their movements are mainly determined by sea currents and waves. It would also seem sensible to assume that the denser the population of settling barnacles on a rock the more likely it is that competition is going to be important. There are data to support this idea for acorn barnacles growing on experimental panels on the pier at Millport Field Centre. Hills and Thomason (2002) found that at higher densities the survival of settled barnacles was lower, and also that they tended to produce fewer young at high densities.

Connell's PhD project was to look at intraspecific competition in acorn barnacles, and the role of predation by Dog Whelks (Connell 1961a). However, in September 1952 he attended a field course in Oxford, run by Charles Elton's Bureau of Animal Population. These influential field courses ran from 1948 to 1956, being attended by research students from around the country and occasionally also by students from overseas who were visiting Oxford as part of their training (Crowcroft 1991). Connell credited this course for introducing him to then-current ideas on interspecific competition. The coursework was field-based in and around Wytham Woods, working on invertebrates (e.g. grasshopper populations) and trapping voles to study their population ecology.

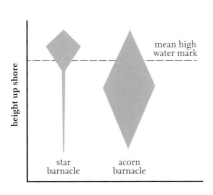

FIG 86. Summary of typical vertical distribution of star and acorn barnacles on the rocks at Farland Point, Millport. The width of the polygon gives a rough indication of how common that species is at that point on the rocky shore. High tides reach the barnacles high on the shore (top of the graph), with star barnacles doing best at a height where they are not covered by every tide. See Box 1 for discussion of the species concerned and their scientific names. Based on Connell (1961b), Gallagher et al. (2015), and personal observations (2019).

BOX 1. Changing scientific names

Academics (such as me) frequently insist that their students use scientific names, rather than just the English ones, for species. The idea is that they are a lot more exact and unambiguous. This is often true – for example, to a North American naturalist 'Red Squirrel' means *Tamiasciurus hudsonicus* while to a European naturalist Red Squirrel is a different species, *Sciurus vulgaris*. Just writing Red Squirrel not only leaves it ambiguous as to which species you are referring to, it could in this case apply to animals in two different genera.

As most readers will be aware, in these two-part scientific names the first word (always starting with an uppercase letter) is the genus, while the second (no uppercase) is the specific name. The usual analogy is with human names: the genus is like a family name (in my case 'Wilkinson') that lumps together several closely related individuals, while the specific name is like a given name that identifies a particular person ('David' in my case) – or in the case of scientific names the particular species.

The idea is that all the species in a particular genus are more closely related to each other than they are to species in a different genus. This is the key point: the scientific name is not just a label to unambiguously identify a species, it is also a hypothesis about evolutionary relationships. So as ideas about relationships are improved it's not uncommon for species to move between genera. In addition, further research can sometimes show that what was thought to be a single species is actually two or more very similar-looking species, again leading to name changes. Such changes have affected a number of the species described in this book, but are particularly apparent in this chapter as they affect all of the key species studied by Connell at Millport. If you read Connell's research papers from the early 1960s (and they are classics that are still worth reading) you will find him using different names than those in current use for all of the main species in his study.

Here is how things have changed. The scientific name for the star barnacles at Millport was *Chthamalus stellatus* when Connell was writing, but they are now thought to be members of the very similar species *Chthamalus montagui*. The scientific name of the acorn barnacles has also changed – from *Balanus balanoides* in Connell's time to *Semibalanus balanoides*. The non-native barnacle now called *Austrominius modestus* was in the genus *Elminius* when Connell found a single individual on the island in the early 1950s. The predatory Dog Whelk is called *Thais lapillus* in Connell's reports but is now *Nucella lapillus*. There have been other name changes on the Millport intertidal rocks too. For example, the black tar-like lichens used to be in the genus *Verrucaria* but are now mainly in the genus *Hydropunctaria*. In truth it can sometimes be hard to keep up with all the changes, especially for groups of organisms that you are not a specialist on!

continued overleaf

Changing scientific names *continued*

Occasionally both the generic and specific names can change too. Most wild-flower identification guides, unless very up to date, call Common Ragwort *Senecio jacobaea*, but it is now *Jacobaea vulgaris*. It has been moved to the genus *Jacobaea*, and because under the formal rules for naming plants a species cannot have the same generic and specific name the specific (second) name has had to change too. The zoological rules are different, so for example the Magpie can be *Pica pica* and the Pine Marten *Martes martes*.

Another student on the course that year was John Coulson, who told me that Connell wasn't deeply into competition at that time – but pointed out that in Elton's group, and especially in David Lack's research group (see Chapter 10), which was situated next door to Elton's, there was a big interest in interspecific competition. Fired up with these new ideas, on returning to Scotland, Connell decided to add additional experiments on interspecific competition to his research, eventually leading to an influential research paper in *Ecology* (Connell 1961b). However, his supervisor was of the view that he shouldn't spread himself too thinly, so he kept quiet about these additional experiments and didn't include them in his PhD thesis.

Connell decided to expand his studies to look at competition between the two common barnacle species at Farland Point, acorn and star barnacles. Looking at the barnacles on the rocks, you can find some areas with a mix of the two species, while on other parts of the shore just one dominates (Fig. 87). You might infer that one species is outcompeting the other, but it's also possible they are reacting to other aspects of the environment – such as the time spent out of water at low tide. The best way to try and work out which processes are most important is through an experimental approach. In the 1950s when Connell was working at Millport there were already influential lab experiments looking at competition using organisms such as protozoa or beetles, but experiments in the wild were rare – and obviously it's what happens under natural conditions that we really want to understand.

Connell mapped the position of barnacles at specific locations, and followed changes over time, by tracing the position of each barnacle onto glass sheets. In addition, in some areas he removed acorn barnacles from around the star barnacles and also moved rocks covered with barnacles up and down the shore to move star barnacles into areas where they were not normally found. He also looked at the effects of predation by Dog Whelks (Fig. 88) by excluding them

FIG 87. Star barnacles dominating the upper shore on Farland Point, Millport. Note the 'kite-shaped' plates covering the centre of the barnacle. The acorn barnacle has a more pointed opening.

from some areas with small cages (this is similar to the Cwm Idwal exclosure experiments described in Chapter 2). He showed that in the absence of competition from acorn barnacles the star barnacles could grow anywhere on the shore (although they did less well on the lower shore), but with competition they were largely restricted to the upper shore where conditions were too difficult for the acorn barnacles to grow. The acorn barnacles were better competitors than the star barnacles, in part at least because they grew more quickly. In addition, the Dog Whelks seemed to preferentially target the acorn barnacles, as they preferred larger barnacles as food, so reducing the extent of the competition between the two species. This is an interesting result, an example of the idea of apparent competition mentioned above, and I will return to ideas on the effects of predation on competition in the next section.

Although this work was carried out in the first half of the 1950s it wasn't published until 1961. By this point the extent and importance of interspecific competition was a major question in ecology, making these apparently conclusive

FIG 88. In the 1950s Dog Whelks *Nucella lapillus* 'were common in this locality' (Connell 1961b). They still are, and this photograph was taken on Farland Point in 2019.

experimental results very important. However, in a nice illustration of the complexity of ecology, when Connell returned to the USA he tried to replicate these experiments on the coast north of Seattle and got less clear-cut results – causing him to become sceptical about the role of interspecific competition in structuring ecological communities (Grodwohl *et al.* 2018). Another interesting aspect of Connell's classic paper in *Ecology*, which looks distinctly odd to modern eyes, is the very limited use he made of formal statistics in analysing his data (for the mathematically inclined: he used only one Spearman's Rank correlation and one Mann-Whitney U test in the entire paper – I am old enough to remember that such rank-based statistical tests were quite fiddly to calculate until the widespread use of computers came in). Today journal editors would require much more statistical analysis of the results to support the study's conclusions, something that became much commoner in ecology over the two decades following Connell's 1961 papers. By the time I was a student in the early to mid-1980s we were being taught lots of statistics for use in analysing the complexity of ecological data sets, something that has grown even more over recent decades, supported by the ever-increasing power of computers.

There is an interesting additional aspect to Joe Connell's work at Millport, related to invasive introduced species. During his PhD studies he found a single individual of the southern-hemisphere barnacle *Austrominius modestus* in 1955. This species was first described by Charles Darwin, based on individuals collected from Australia and New Zealand, and was accidentally introduced to Britain around the 1940s (the first record being from Chichester Harbour on the south coast of England, around 1943). It wasn't recorded again on the Isle of Cumbrae until 1959, despite the area being well studied as a major site for marine biology research. It's now much more common on the island, although still relatively rare at Farland Point (Gallagher *et al.* 2015; personal observations 2019). It tends to occupy the same part of the shore as star barnacles, and on Cumbrae work by Mary Gallagher and colleagues suggests it's most common at sites with the fewest star barnacles; often these are sites with a lot of human disturbance, for example around jetties. Competition between native species and introduced ones is a frequent cause of concern for conservation, and the squirrels discussed in the next section provide a classic non-marine example.

COMPETITION AND CONSERVATION – RED AND GREY SQUIRRELS

Think of competition in the context of British natural history, and one of the examples that would occur to many people is the effects of introduced Grey Squirrels *Sciurus carolinensis* on the native Red Squirrel *S. vulgaris*. The story of Grey Squirrels in Britain is well known. Between 1876 and 1929 they were introduced from North America to over 30 sites in Britain, and from these have spread to be found in most of England and Wales, along with parts of Scotland and Ireland (Fig. 89). As the Grey Squirrel spread, the Red became extinct across much of the country. Summarising the position in the early 1950s, Monica Shorten (1954) wrote that 'The general impression is that the introduced species is exerting some unfavourable influence on the native squirrel, but we still do not know how it is doing this.' For much of the second half of the twentieth century the explanations tended to focus on scramble competition between the two species, with Greys outcompeting Reds for food. For a long time it had been suggested that the Grey Squirrels may have infected the Reds with a disease, although this tended to be considered at best a minor issue (Shorten 1954). However, over the last couple of decades it has become clear that squirrelpox virus (SQPV) has played an important role in the replacement of Reds by Greys (Tompkins *et al.* 2003).

FIG 89. The Grey Squirrel *Sciurus carolinensis* is native to parts of North America, but was introduced to Britain repeatedly during the nineteenth and early twentieth centuries. The earliest successful introduction appears to have been at Henbury, Cheshire, in 1876, although there are occasional accounts of what appear to be Grey Squirrels in Britain as far back as the 1820s. Following the Henbury introduction there were at least another 30 introductions up to the end of the 1920s. The release of squirrels into the wild was banned in 1938; in the mid-twentieth century free cartridges were supplied to squirrel shooting clubs, and a bounty paid on dead Grey Squirrels, but by this point the Greys were too well established for such measures to succeed in eliminating them (Shorten 1954, Lever 1977).

Today in Britain we tend to associate Red Squirrels with coniferous woodland, and Grey Squirrels with deciduous woodland. However, before the arrival of the Greys, Reds were widely found in deciduous woods, with Hazel *Corylus avellana* nuts being a particularly important food in this habitat. Indeed, in the eighteenth century Gilbert White described Red Squirrels as living 'much on hazel nuts' (see Chapter 8 for more on White and his observations). Importantly acorns, while good Grey Squirrel food, are poor food for Reds. Acorns are chemically protected by polyphenols (such as tannins) which Red Squirrels are thought to find difficult to digest. The ability of the Greys to make good use of acorn crops, especially in good acorn years, seems to put Reds at a competitive disadvantage against Greys in deciduous woodland. One potential problem with the key study that helped established these conclusions, by comparing squirrel populations in deciduous and coniferous woodland in England, also serves to highlight how rare Red Squirrels now are in British deciduous woodlands (Kenward *et al.* 1998). The data on Reds in deciduous woodland was based on populations in only three woodlands, all on the Isle of Wight, which was free from Greys and so still had surviving Reds. In most cases Reds become extinct within a maximum of 10–20 years of Greys arriving in an area – unless there is management to try and keep the Greys at low population levels.

The very rapid replacement of Red Squirrels by Greys seemed too fast to be just down to competition for food, and at the start of the twenty-first century this led to renewed interest in the possible role of apparent competition due to disease. Attempts to construct mathematical models of the decline of Red Squirrels in Norfolk suggested that the effects of disease needed to be included to achieve the rate of decline seen in the actual data (Tompkins *et al.* 2003).

The effects of squirrelpox are illustrated by the Red Squirrel populations on the Sefton coast, just north of Liverpool. Along the coast Red Squirrels inhabit coniferous woodland, mainly at the back of sand-dune systems, but some are also found in scattered woodlands further inland. These plantations were mainly created in the late nineteenth century, in part to try and reduce problems with sand blowing from the dunes onto the surrounding area. This suggests that the Red Squirrel population here may be relatively recent, and there is evidence that the ancestors of at least some of these squirrels were introductions from continental Europe in the 1940s. Monica Shorten (1954) made no mention of Red Squirrels here in her mid-twentieth-century review of British squirrel biology, but they were common by the late 1960s and early 1970s when I was a child, visiting on days out from Manchester. Around the start of the twenty-first century there were somewhere between 650 and 900 breeding adult Red Squirrels in the coastal pine woods, with another 300–400 in adjacent urban areas and small inland woodlands – but Grey Squirrels were increasingly being reported from the area. In 2003 squirrelpox was first reported from the Sefton coast Red Squirrel populations, and by 2008 there had been a decline of 80–90 per cent in one of the best-known Red Squirrel populations in the area, at Lifeboat Road, Formby (Smith 2009).

This outbreak of squirrelpox has been studied in detail by a team from the University of Liverpool and the Lancashire Wildlife Trust (Chantrey *et al.* 2014). Transect surveys of squirrels in the woodlands along the coast showed substantial declines in Red Squirrels associated with the squirrelpox outbreak. In addition, 448 dead animals were collected, of which 151 had died directly as a result of the virus (with another 33 infected Red Squirrels which had died from other causes). Several interesting conclusions followed from this work. First, the virus was clearly the major cause of the decline in Red Squirrels, while comparison of densities of Red and Grey Squirrels provided no evidence of other, non-disease-related, competition. Note that the major habitats here were coniferous plantations, rather than the deciduous woodland from which there is better evidence for food-related competition between these two squirrel species. Second, the testing of Red Squirrels caught alive showed that some of them, although healthy, carried antibodies against squirrelpox. This was an optimistic result, showing that Red Squirrels in the wild can recover from a virus which had previously been assumed to be invariably fatal.

The history of the impact of Grey Squirrels on the native Reds in Britain has been called 'one of the best-documented examples of the negative impact induced by biological invasions on native ecosystems' (Romeo *et al.* 2019). One of the reasons it's so well documented is that in the 1940s and 1950s a large number of amateur naturalists across the country diligently recorded what was going on with their local squirrels. Their efforts were marshalled by Monica Shorten, who was building on similar work carried out by Doug Middleton in the 1930s (both encouraged and advised by Charles Elton). Clearly citizen science in Britain has a history which greatly pre-dates the rise of the internet.

Recently much of the emphasis has been on apparent competition mediated by squirrelpox. In the coniferous plantations of the Sefton coast this seems much more important than scramble (exploitation) competition for food or other resources. However, at least in deciduous woodland, scramble competition seems important too. The situation in Italy forms a useful comparison here. There are now expanding introduced Grey Squirrel populations in several parts of northwestern and central Italy. These expanding Greys are associated with declining Red Squirrel numbers, but recent research has failed to find any convincing evidence for squirrelpox in the northwest of the country, which is where most of the Greys are currently found (Romeo *et al.* 2019). In this case it appears to be entirely scramble competition that is leading to the decline of the Reds.

The concept of apparent competition doesn't apply just to diseases (microparasites, such as viruses and bacteria) but also to the effect of predators, as with the Dog Whelks in Connell's studies at Millport. In the case of the squirrels, the recovery of the Pine Marten *Martes martes* has recently been attracting a lot of interest in both Britain and Ireland. The widespread decline and local extinction of the Pine Marten was particularly associated with the development of sporting estates during the nineteenth century. By 1915 the British population was largely restricted to the Lake District, north Wales and northwest Scotland. Prior to the nineteenth-century rise of gamekeeping there had been earlier declines, in part associated with the loss of suitable habitat (Langley & Yalden 1977). The Pine Marten in Ireland had a similar history of decline. However, over my lifetime the Pine Marten has started to expand again; for example, Katherine Sainsbury and colleagues (2019) estimate that between 1975 and 2015 the Scottish population expanded at a rate of around 1.7 km per year, and more recently there have also been expansions south of the Scottish border. The Pine Marten eats a wide range of foods, including small mammals – such as mice and voles – squirrels, birds and their eggs along with carrion, fruit, nuts and fungi. Importantly for the central concerns of this chapter, much of its foraging is on the ground – although it can take birds and squirrels in the tree canopy too.

Apparent effects of the increase in marten numbers on competition between the two squirrel species were first described from Ireland, then confirmed and extended by studies in Scotland (Sheehy *et al.* 2018). I will focus on this later Scottish study here. Emma Sheehy and colleagues used a mix of baited glue traps (which collect hair samples from visiting animals, which can then be identified to species and genetically analysed to identify individual martens) and camera traps at three locations across Scotland. These data were subjected to statistical analysis to look at the interaction between the two squirrel species and Pine Martens. The key conclusion was that the presence of Pine Martens had a negative effect on Grey Squirrels and an 'indirect positive effect' on the Reds (Fig. 90); hence the title of the research paper by Sheehy and colleagues: 'The enemy of my enemy is my friend'.

But what is it about Grey Squirrels that makes them more susceptible to Pine Marten predation than Reds? One possibility that has been suggested is that the heavier Greys spend more time on the ground than the Reds, making them more easily caught. However, the Scottish study also provided evidence that the Greys may be less cautious than Reds in areas with high Pine Marten numbers. In the study Greys continued to regularly visit feeders in high-density marten locations, while the Reds showed more caution. The suggestion is that because the Greys did not evolve alongside Pine Martens they are lacking adaptations to avoid them. If this is correct, it suggests that the beneficial effects of Pine Martens on Red Squirrel populations may be short-lived, if the Greys rapidly evolve to become

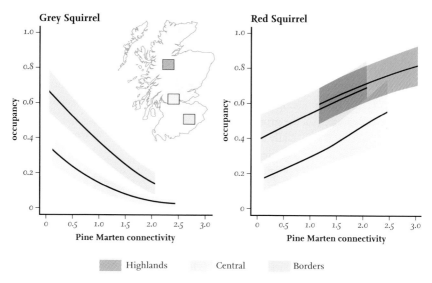

FIG 90. Predicted relationship between occupancy by Grey (left panel) and Red Squirrels (right panel) in relation to Pine Marten connectivity – a measure of the frequency of use of a location by Pine Martens. These relationships were calculated by Sheehy *et al.* (2018), based on surveys described in the text. (Licensed under CC BY 4.0)

more cautious. However, in their native range Grey Squirrels live alongside a couple of other marten species (while many Reds have for generations lived in areas with no martens), making such an explanation seem unlikely to be the whole story.

This chapter has so far used two principal examples in discussing competition, barnacles on a rocky shore and squirrels in woodland. These superficially appear so different that it's not obvious that explanations applying to one would be particularly useful if applied to the other. However, in both cases the role of a predator – Dog Whelk or Pine Marten – appears important. Indeed, in the squirrel case the marten-mediated apparent competition has attracted a lot of interest from conservationists, as it may prove to be important for Red Squirrel conservation in the future if martens continue to spread.

Are these two cases exceptional, or do predators usually, or even always, reduce the extent of competition between their prey species? Clearly in ecology we would like to find relationships that apply across all types of organisms and different habitats, but we need to consider the possibility that the world may be more complex than we would sometimes like it to be. So how general is this apparent role of predators in weakening competition? Although it's often assumed that predation reduces competition, there are also examples of it increasing competition, or having little effect. However, in cases where the predator is selectively targeting the superior competitor, then a reduction of competition would be expected (Chase *et al.* 2002). To understand the difficulties, imagine the example of competition between Red and Grey Squirrels and the effect of introducing an arboreal predator (remember Pine Martens preferentially forage on the ground). If the predator mainly hunted in trees then it would tend to preferentially target Reds – increasing the effects of competition between the two squirrel species, rather than reducing it. So the interaction of these two important ecological processes of predation and interspecific competition can give a range of different results depending on the specifics of the system being studied.

INTRASPECIFIC COMPETITION – THE EXAMPLE OF SPERM COMPETITION

This chapter, so far, has concentrated on interspecific competition, that is competition between members of different species. However, members of the same species also compete – indeed, as described above, it is often assumed that such competition may be particularly acute, given that members of the same species are likely to have very similar resource requirements. Intraspecific competition can be for the same sorts of resources as interspecific competition, and for familiar animals such as garden birds these resources include food and nest sites. However, for sexually reproducing animals such as birds there can also be competition for mates, a resource not normally the subject of interspecific competition. This makes various aspects of mate choice a useful introduction to intraspecific competition.

Consider a cowpat. Much cattle dung in Britain is now biologically dull due to various veterinary products poisoning many of the organisms that should colonise the dung. However, the dung of organically reared cows makes a fascinating mini-ecosystem. Here fungi grow from the cowpats (Fig. 43), the specialist water beetle *Sphaeridium scarabaeoides* tunnels and 'swims' through the more watery dung, while Yellow Dung Flies copulate on the surface (Fig. 91). These dung flies played an important role in the history of ideas on intraspecific competition. Lie down in the grass, watch the flies, and one aspect of competition is easily observable. The males physically compete for females and can sometimes be seen displacing a rival during copulation. During the 1960s the reproductive behaviour of these flies was the subject of a study by Geoff Parker, who started

FIG 91. Yellow Dung Flies *Scathophaga stercoraria* copulating and mate guarding on cow dung. Note the much larger size of the males. The fact that the males often physically fight for access to females is the likely explanation for male size – smaller males struggle to get access to females.

work on the insects while still an undergraduate at the University of Bristol and continued it for his PhD. He was able to show that competition for mating went beyond the physical tussles you can watch on the cowpat, and continued after copulation was completed. Many of the female flies mate with multiple males, and competition between the sperm of rival males to fertilise the female's eggs continues inside her reproductive tract. By experimentally sterilising the sperm of some males (so that although it could still fertilise an egg an embryo couldn't develop), Parker was able to show that most eggs were fertilised by the last male to copulate with the female. This makes it important for the first male to try and avoid any other males copulating with the female, and they try and stay on top of her until she has laid her eggs – as seen in Figure 91. So there is both competition to copulate and also potentially competition between rival sperm after copulation if the mate-guarding strategy fails to prevent further copulations.

Geoff Parker published a series of research papers on his dung fly work, but it was a 1970 review paper that has come to be seen as particularly influential, effectively founding sperm competition as a research area (although the term was first used in a study of fish in the 1930s). In this work he summarised his dung fly results, along with information on similar phenomena in other insect species, and developed the idea that intraspecific competition amongst sperm might be an important aspect of the ecology of many animals. This work has led to Parker being called 'indisputably the father of sperm competition' (Birkhead 2000).

Moving from insects to more familiar vertebrates, the relationship between body mass and testes mass suggests that some interesting biology is involved – but to appreciate this a bit of background natural history is useful. Many bird species appear to be monogamous, the same pair staying together for the breeding season or even longer. For example, in his classic mid-twentieth-century study of the behaviour and ecology of British Robins, David Lack (1946) wrote that 'In robin pairs constancy for the breeding season is the rule' (Fig. 92). Modern genetic methods, which allow paternity testing of chicks, show that this is not completely correct. It has been demonstrated that a proportion of the chicks of many apparently monogamous birds have been fathered by birds other than the male associated with the pair. Because of this, the term often used today is 'social monogamy' – so the system appears monogamous even if some of the chicks come from so-called extra-pair copulations. As his biographer has pointed out, with the benefit of hindsight we can see evidence for social monogamy in some of David Lack's Robin observations (Anderson 2013). At the time of his Robin studies Lack was working as a schoolmaster at Dartington Hall School, in Devon. As well as studying the Robins in the school grounds he also had Robins in aviaries, and had seen a male captive Robin copulating with a female that wasn't his mate – however, Lack was understandably sceptical of the importance of observations of birds in captivity, so made nothing of the observation. It wasn't until the 1980s that genetic fingerprinting technology allowed attempts to

FIG 92. The Robin *Erithacus rubecula* is one of many British bird species classified as socially monogamous. As described in the text, the Robin was the subject of classic studies of its behaviour and ecology by David Lack during the second half of the 1930s.

quantify the extent of extra-pair copulations in socially monogamous birds, with some of the first studies being on House Sparrow *Passer domesticus* populations near Nottingham. These were described in two reports in the same issue of the journal *Nature*, and showed mixed paternity for chicks in the same nest (Burke & Bruford 1987, Wetton *et al.* 1987).

Returning to the relation between testes size and body mass, a recent study assembled data on 1,913 different vertebrate species, showing quite a spread in the data (Fig. 93). A simple guess at what such data should show is that the graph should be a nice straight line, with larger species simply having proportionately larger testes. Clearly there is in fact a lot of variation away from this simple pattern, with some species having relatively small testes for their size while others are much better endowed, which explains why all the data do not neatly lie along a straight line in the graph. Since the early work by Parker, the assumption has been that in animals with sperm competition you would expect larger testes size, as that allows them to make more sperm, so aiding them in competing against the sperm of other males (and also to outwit attempts by the female to manipulate which sperm fertilises her eggs). Apart from assembling the largest ever data set of vertebrate body size and testes size, the key novelty of this study was that it was able to compare these data with what is known of the animals' evolutionary history and their mating system. The results for birds were particularly interesting, with repeated cases of testes size reduction evolving in socially monogamous bird species – although interestingly the researchers couldn't find evidence for a similar reduction in testes size in other apparently monogamous vertebrates. So although extra-pair copulations mean that there is scope for sperm competition in socially monogamous birds, it appears that it's not extensive enough to select for large testes size, and the savings in resources from having smaller testes outweigh any disadvantages from reduced sperm production. One possibility with birds is that flight is energetically expensive, so making any opportunity to save energy elsewhere especially important.

This idea of sperm competition raises an obvious question: what's in it for the female? From an evolutionary point of view, why should she be interested in being mated by more than one male? There are a number of possible answers to this question, and the actual explanation likely varies from case to case (Davies *et al.* 2012). One possibility is that it's sometimes outside the female's control. In the case of Yellow Dung Flies the males are much bigger than the females, as can be seen in Figure 91, and in soft cowpats the female can be pushed into the dung and drowned as males fight over her. In this case she likely has little scope to choose whom she mates with. However, since birds can fly, and few male birds have a penis, the female has much more choice with whom she mates,

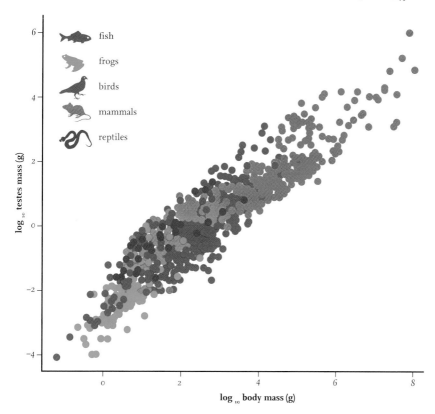

FIG 93. Testes mass and body mass for 1,913 species of vertebrate, from Baker *et al.* (2020). Note the scatter of points: while there is an obvious tendency for larger animals to have larger testes, these data show quite a lot of variation (note also that log scales have been used to graph these data, which makes the data fall along a straight line and the graph easier to interpret). (Licensed under CC BY 4.0)

as it's hard for a male to force a copulation – so it seems likely that she may be actively choosing to mate with multiple males. There is a range of potential direct benefits to a female from multiple matings, the most obvious being that more sperm may help ensure fertility. In addition, it may be to her advantage that several local males believe that her offspring might be theirs, as they may all help her rear them – as can be the case with Dunnocks *Prunella modularis*.

There are potentially more indirect benefits too. If your offspring have multiple fathers then they will be genetically more diverse than if they were all fathered by the same individual – this may for example increase the chance of at least some of them having genes that give resistance to a particular disease. It may seem obvious that such ideas would be applicable only to intraspecific competition in sexually reproducing animals. However, you can apply very similar ideas to plants, with pollen grains from multiple males landing on the female parts of flowers, and potentially the chance for the flower to bias which pollen actually fertilises it. Indeed, in 1983 Mary Willson and Nancy Burley wrote an important and provocative book entitled *Mate Choice in Plants*, applying these apparently animal-focused ideas to botany. These ideas may be more widely applicable than they appear at first sight, and not only of interest to zoologists.

CONCLUDING REMARKS

Competition, simply defined as two individuals attempting to use the same limited resource to the detriment of them both, is a key idea in ecology. Although I mainly use animal examples in this chapter, these ideas are just as applicable to other groups – from microscopic bacteria to large oak trees. Competition is an idea that is applied in many of the chapters in this book; in particular, for example, I discuss the importance of competition in the context of competitive exclusion in the chapter on ecological niches (Chapter 10).

In the current chapter I have introduced the idea of competition and the various ways that it has been classified: inter- and intraspecific, scramble, contest and apparent competition. Apparent competition, due to the involvement of a predator or parasite, may sometimes be important in reducing the effects of competition; however, the effectiveness of this can vary from case to case. In the two extended examples of interspecific competition used in this chapter – barnacles at Millport and squirrels in Britain and Ireland – a predator does apparently reduce the extent of competition, making it more likely that both species can survive at the same location. Intraspecific competition (that is, competition between members of the same species) can be for the same types of resources as those competed for between members of different species. However, intraspecific competition can also be for mating opportunities, and such cases – especially that of sperm competition – provided the main examples used in this chapter. Since the classic work by Geoff Parker at the start of the 1970s this has become a major research area for ecologists with a particular interest in behaviour and evolution. Although competition features very prominently both in this book and in ecological theory in general, there are also relationships between different species which appear mutually beneficial. These are the subject of the next chapter.

CHAPTER 7

Cooperation in the Cairngorms

The mountainside was dark and the light from my head torch illuminated a small patch of snow in front of me as I slogged uphill, progress marked by occasionally passing the supporting towers of chair lift and ski tows. It was December 1992, and we were on our way to a remote climb in the Cairngorms. On a short northern winter day, all the daylight is needed for climbing and is not to be wasted on the walk in to the crag, hence the predawn start. The sun started to rise as we were crossing the snowy plateau heading towards the top of Glen Avon, with the red light of dawn slowly revealing the closest landscape that Britain has to arctic wilderness.

Over a quarter of a century later, it's September 2018 and I am again toiling up the slopes of Coire Cas, this time in daylight. The chair lift has long gone, replaced by a funicular railway, which is itself now in need of replacement or very expensive repairs. Although the day is hardly wild by Cairngorm standards, up high the wind still makes itself felt. The weather map showed the jet stream blowing right across the Cairngorms, albeit at a much higher altitude than the 1,245 m summit of Cairn Gorm itself. Showers blow through on the wind, propelling cold drops of water with a force that makes you check to see if they are hail as they sting your face – the gusts threatening to blow you over if they catch you off balance as you negotiate the rocky ground.

Conditions on the Cairngorm plateau are tough, at least when viewed by human standards. Even in summer it can be distinctly challenging, and in winter it's often fully arctic (Fig. 94). Anything living up here year-round needs to be able to survive these conditions – but if you can do so then there are benefits, as you escape competition from the many organisms that just can't live here. The vegetation is low and patchy – the ground best described

FIG 94. The Cairngorm plateau in winter, the nearest to arctic habitats that Britain can provide.

FIG 95. Boulder above Coire an t-Sneachda in the Cairngorms. The bright green patches are the Map Lichen *Rhizocarpon geographicum* (see Fig. 4 for a close-up of this species).

as a fell-field, with barren stony bare ground between patches of alpine heath vegetation (Ratcliffe 1981, 1991). This is a vegetation often associated with polar regions, and the bare ground is a testament to the conditions; here, erosion is a visibly active process, not some bookish concept from a geography text. Long-lasting snow beds add to the arctic character of the vegetation, and provide habitat for rare plants and lichens.

Some of the bare rock is genuinely devoid of life, at least life of a size visible with a hand lens; the wind that's threatening to unbalance the mountaineer also sand-blasts the more exposed rock surfaces, making them almost uninhabitable. However, many rocks are far from lifeless when examined more closely. Most obvious are the almost fluorescent bright green patches of Map Lichen splashed across the boulders as if by some abstract artist (Fig. 95), and once you get your eye in it's obvious that plenty of other lichens can survive up here. Lie prone out of the wind and examine sheltered hollows between rocks and there are the twisted straps of Iceland Moss (despite its name, this is another lichen), and amongst the alpine heath vegetation is the macaroni-like lichen *Thamnolia vermicularis*, sometimes called the white worm lichen (Figs. 96 & 97).

FIG 96. Iceland Moss *Cetraria islandica* in a sheltered site between rocks on the Cairngorm plateau. Despite the common name this is a lichen, not a moss. The strap-like lobes are just under 1 cm wide. The rarer, smaller *C. ericetorum* is also found on parts of the plateau (Gilbert 2000).

FIG 97. The white tubes of *Thamnolia vermicularis* growing amongst Heather *Calluna vulgaris* high in the Cairngorms. It can be common on the bare gravelly ground between Heather bushes (Gilbert 2000). A patch of another lichen (*Cladonia* sp.) can be seen to the right of the photograph.

Lichens can survive not only here in the Cairngorms but in some of the most inhospitable places on Earth – from high on Himalayan giants that dwarf these Scottish mountains to continental Antarctica, from hot deserts to the vacuum of space as part of experiments to investigate possibilities of life on other planets (Stephenson 2010). Despite having a single scientific name, lichens are in fact composite organisms. Such apparently cooperative relationships are common in biology, although historically somewhat overlooked by many ecologists. Composite organisms form the main subject of this chapter, and lichens are a good introduction to the topic. However, before thinking about lichens in more detail it's worth considering why such organisms have often seemed strange to biologists.

THE 'PROBLEM' OF COOPERATION

There are many examples of cooperation in British plants which appear straightforward to understand. Consider the many species which can grow clonally via stolons – above-ground stems growing along the ground and giving rise to new plants (these are often called 'runners' when the stem is long, although this is a rather loose distinction). The same thing can happen with underground stems, where they are termed rhizomes (Bell 1991). Familiar British examples of this type of growth include White Clover *Trifolium repens*, Wild Strawberry *Fragaria vesca* and Bracken (Fig. 98); globally, around 35 per cent of plants can be linked by such connections (Vannier *et al.* 2019). The individual plants that arise from these stolons and rhizomes are genetically identical, give

FIG 98. Bracken *Pteridium aquilinum* covering a hillside at Crowden in the northern Peak District. Bracken largely spreads by rhizomes, forming large clonal patches.

or take the occasional mutation, and so it's not surprising that they behave cooperatively, for example by sharing nutrients (Fig. 99). In human terms this is like cooperating with your identical twin.

The examples of apparent cooperation in composite organisms discussed in this chapter are very different. They are not examples of cooperation between very close relatives with most of their genes in common; indeed, they are not even examples of cooperation between more distantly related members of the same species. Instead they are interactions between members of different species – such as fungi and algae in the case of lichens. Perhaps the most familiar example to gardeners are the nitrogen-fixing bacteria in the roots of legumes such as peas.

Why is cooperation of this sort seen as a problem? During the 1960s and 1970s the idea that thinking about evolution at the gene level was the way to understand organisms – including the evolution of cooperation – became widespread in biology. This is now the standard way of thinking, and it became widely known following its popularisation in Richard Dawkins' 1976 book *The Selfish Gene*. If you take a gene's-eye view of things then it can make good evolutionary sense to help close relatives that have more genes in common with you than would a member of the general population. By helping them you are furthering the interests of your own genes.

Clonal growth

clonal fragment

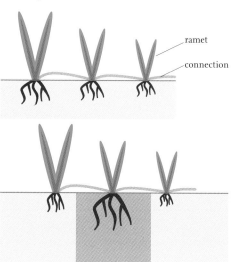

ramet

connection

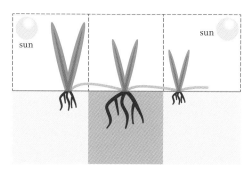

sun

sun

Division of labour

FIG 99. Clonal plant linked by connections which can, as here, be above ground (stolons or runners) or below ground (rhizomes). The individual genetically identical 'plants' are referred to as ramets. Nutrients, and other resources, can move between these ramets. In the lower two illustrations more nutrient-rich soils are shaded brown. In the middle panel the central ramet in the nutrient-rich soil has the possibility of subsidising the other ramets in less nutrient-rich soils, while in the lower panel a division of labour is possible, with the central ramet having more access to soil nutrients but the two outer ramets having more access to light for photosynthesis. Modified from Vannier *et al.* (2019). (Licensed under CC BY 4.0)

These ideas were mainly developed in the 1960s by scientists such as Bill (W. D.) Hamilton in Britain and George Williams in the USA. However, as so often in science the basic idea emerged slowly, and various people had in fact been starting to think about this gene-level approach before the 1960s. One of these scientists was J. B. S. Haldane, and there is an oft-told story of Haldane sitting in the long-gone Orange Tree pub on Euston Road in London around the early 1950s talking evolution over a beer or two. To illustrate the point he was making he imagined the evolutionary implications of risking your life to save drowning close relatives (who share many genes in common with you), and suggested that he would sacrifice his life to save at least two siblings or eight first cousins. The key point here is that a sibling has a probability of 1/2 of having the same form of any particular gene that you have, while the probability is 1/8 for a cousin (Sherratt & Wilkinson 2009). Taking such a gene-level approach, it's easy to understand cooperation between two genetically identical 'individuals' (ramets; Fig. 99) in a clonal plant, but hard to see how cooperation can evolve between two or more very different species, as happens in a lichen. In such a case the cooperative behaviour isn't obviously helping to get your genes into future generations. Later in this chapter I will return to some of the ways in which such associations may evolve.

Another problem is what to call an apparently cooperative interaction between two organisms. The term most widely used in popular science writing and TV nature documentaries is 'symbiosis', but this is a confusing word used in several different ways in ecology. Symbiosis is a nineteenth-century term originally coined to describe any close, long-term association between different organisms – so it applied to parasitism as well as to more cooperative interactions. This is still the most common usage in the technical literature. However, by the early twentieth century symbiosis was sometimes being used as a term purely for mutually beneficial interactions, while still being used by others in its original sense. Confusingly, both uses of symbiosis can be found in recent texts (Wilkinson 2001). For clarity the alternative term 'mutualism' seems the better one to use when describing cooperative associations such as those discussed in this chapter. Symbiosis is better restricted to its original meaning of any relationship between species, where the association is so close that it's not immediately obvious that more than one organism is involved. In this sense mutualisms can be either symbiotic, as in the lichens discussed in the next section, or non-symbiotic, where different species are clearly involved in a cooperative interaction. The main focus of this chapter is on symbiotic mutualisms, which are very important in many ecosystem processes. However, non-symbiotic ones such as insect pollination or seed dispersal by birds are also of fundamental importance – indeed, some of my own research has been on the dispersal of seeds by birds, such as the waterbirds described at the start of Chapter 5.

LICHENS – AN INTRODUCTION TO MUTUALISMS

Below the arctic plateau of the Cairngorms are forests dominated by Scots Pine *Pinus sylvestris* (Fig. 100). With their mix of open heath with scattered trees and more

FIG 100. The edge of Rothiemurchus in the Cairngorms, on the track from Loch an Eilein to the Cairngorm Club footbridge.

densely wooded areas, these are very different from the more domestic southern woodlands such as Wytham near Oxford. As Derek Ratcliffe (1981) wrote, 'There is here, at least in those parts where human activity is little in evidence, the quality of the real wilderness of the boreal regions.' There are lichens here too, but in the somewhat more sheltered conditions you can find larger types than those on the high tops. Most obvious are the beard lichens hanging from the branches of the trees, a collection of confusingly similar species in the genus *Usnea* (Fig. 101). The hair-like strands of Grey Horsehair Lichen *Bryoria capillaris* also decorate tree branches here – in Britain it's a species largely restricted to these Caledonian forests of Scotland.

But what are lichens? It's not immediately obvious. Look at the photographs of lichens in this book, or even better go outside and look at some living examples. If you knew nothing about lichens, what would you guess? They are superficially plant-like but obviously rather different from a typical plant. Their nature was the subject of significant controversy in nineteenth-century science; however, by around 1880 scientific opinion in Britain was coalescing around the idea that they were dual organisms composed of a fungus and photosynthetic algae. A simple definition of a lichen is that it's a stable self-supporting consistently identifiable association between a fungus and photosynthetic microorganisms – these are usually green algae but can be cyanobacteria (Fig. 102). As composite organisms it's not obvious how you give them a single

FIG 101. Lichens in the Rothiemurchus Forest, including a prominent beard lichen *Usnea* sp. and *Hypogymnia physodes* (to the right of the photo). While *H. physodes* can survive in relatively high levels of sulphur dioxide pollution, the beard lichens require clean air (see Chapter 12 for more on lichens and air pollution).

FIG 102. The dog lichen *Peltigera membranacea* growing at the edge of Abernethy Forest in the Cairngorms (the common name 'dog lichen' applies to a number of similar species in this genus). This is one of the commonest large lichens in Britain and has cyanobacteria as its associated photosynthetic microorganism. The majority of lichens have green algae as their photosynthetic microbes, and in a few cases both algae and cyanobacteria are present. The study by Knowles *et al.* (2006) in Minnesota described in the text used several species of *Peltigera* (all species also found in Britain) but not *P. membranacea*.

scientific name. The solution used, which became the consensus approach in the mid-twentieth century, is that the name actually applies to the fungus, which makes up the main body of the lichen (Gilbert 2000, Wilkinson 2018).

Look inside a typical lichen, and under a microscope you see a tangle of spaghetti-like strands (the fungal hyphae) and small pea-like spheres (the algae). Unsurprisingly most of the algae tend to be near the upper surface of the lichen, as this is where most of the light is for photosynthesis (Fig. 103). Lichens are generally interpreted as mutualisms, that is mutually beneficial interactions, where the main fungal body provides a safe protected environment for the algae which in turn provide the fungus with the products of photosynthesis. However, there has been a long-running alternative view of lichens which sees the fungus very much in charge and exploiting the photosynthetic microbes. If you think of human relationships, this confusion about the mutualistic nature of lichens seems less surprising: one can describe many of our own interactions as either cooperative or at least partly exploitative, depending on the level of cynicism of the commentator. For example, the rise of the tech industry in California's Silicon Valley in the 1970s has been explained as at least in part due to cooperation between apparently competing computing firms (take your pick which aspect you emphasise, competition or cooperation, depending on which school of economics you favour), and similar ideas can be applied to the benefits

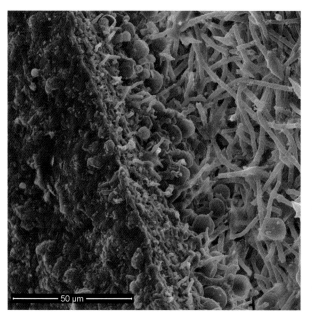

FIG 103. Scanning electron microscope image of the structure of a foliose ('leafy') lichen. The left-hand side of the image shows the upper surface of the lichen, and to the right this has been broken away to show the structure underneath. The spaghetti-like structures are fungal hyphae, while the round objects are the photosynthetic algae. The scale bar is 50 μm long (i.e. 0.05 mm).

of cooperation in ecological communities too (Cohen 1998). Perhaps defining lichens as either 'mutualistic' or 'exploitative' tells you as much about the world view of the scientist in question as it does about the workings of lichens.

The lichens that contain cyanobacteria can obtain another advantage from the photosynthetic microbes, as these can fix nitrogen (see Chapter 5). This not only benefits the lichen as a whole but also the surrounding environment if biologically usable nitrogen compounds leak from the lichen into the soils. This leakage has been shown to be the case for various species of dog lichen (Fig. 102) in forests in Minnesota where soil nitrogen levels were raised for a metre or so around the location of the lichen (Knowles *et al.* 2006). This is an idea of much wider relevance, and the ecological importance of leakage of nitrogen into the environment from cyanobacteria is described in more detail in the context of lakes in Chapter 5. The occurrence of cyanobacteria as well as green algae in lichens goes back deep into geological time. Lichens don't easily fossilise, but the earliest fossils of clearly modern-looking lichens come from rocks on the Welsh border. These are around 415 million years old, and microscopic examination of the fossils shows that some had unicellular ('presumably green') algae, while others contained cyanobacteria (Honegger *et al.* 2013).

The view of a lichen under the microscope will reveal not only the fungus and its photosynthetic partner but a host of other organisms too. A nice example of what can be found is provided by a study of protists and microscopic animals living in lichens on trees around Belfast (Roberts & Zimmer 1990). This 30-year-old study used a simple approach easily replicable by anyone with access to a microscope, but still found 20 different genera of protists, along with nematodes, rotifers, tardigrades, mites and a springtail (the protists were cultured in spring water with a boiled rice grain added to provide nutrients). A more recent study of lichens from Colorado, using the methods of molecular biology, not only found a wide range of protists and microscopic animals, but also a long list of fungal species living within the body of the lichen (Bates *et al.* 2012). Clearly one can potentially view a lichen as a mini-ecosystem – like an oak woodland, but on a very small scale, containing a wide range of organisms (Hawksworth & Grube 2020). However, recent studies, many of them using molecular methods, suggest that things may be more complex, and more interesting than this woodland analogy suggests.

The idea that things may be more complex has a long history. For at least 100 years people have been seeing bacteria when they looked at fragments of lichen under the microscope, and wondering if they might have some role in the lichen mutualism. With modern molecular methods allowing more detailed studies of the ecology of microbes, these long-ignored ideas about the wider importance

FIG 104. Some lichens can be hundreds of years old. The age of patches of lichens growing on rocks can be estimated from their size – using data collected from lichens growing on dated gravestones. In Britain some of the oldest estimated ages are for lichens on the Rollright Stones, a prehistoric stone circle near Oxford. Vanessa Winchester estimated that some of these lichens might have started growing over 800 years ago; in other parts of the world, at high latitude or altitude, some lichens have been estimated to be up to a few thousand years old (Gilbert 2000). The photo shows lichens on one of the Rollright Stones, including two species of *Calopacta* (yellow lichens) and *Solenospora candicans* (white). Note how the centres of some of the *Calopacta* are now missing; nutrients from these old parts are likely to have been recycled to support the actively growing edges – as described in the main text.

of bacteria in lichens are starting to be treated more seriously (Grube & Berg 2009, Wilkinson 2018). It's long been known that cyanobacteria, when present, are an important part of the lichen, but there is now good circumstantial evidence that some of these other bacteria may be playing useful roles too – for example, contributing chemicals such as hormones or antibiotics and helping recycle nutrients within the body of the lichen. As lichens can be very long-lived organisms (Fig. 104), it may be important to release nutrients from old dying parts of the lichen's body for reuse in the more actively growing regions. Studies of the bushy lichen *Cladonia portentosa* by Chris Ellis and colleagues (2005) – using a rare stable isotope of nitrogen as a 'label' (Box 2, page 170) to follow how the nitrogen moved in the body of the lichen – showed that on a timescale of a couple of years nitrogen compounds were recycled and moved from senescent regions to more actively growing ones. Several people have suggested that a role for bacteria in facilitating such processes seems very plausible.

In addition to the bacteria a range of yeast species (yeasts are microscopic fungi) has also recently been found in lichens, and again the thought is they may be contributing useful chemicals – for example, helping to protect the lichen from attack. It's also possible that some of the protists living in lichens may be beneficial, although this is rather more speculative. What we are currently lacking is experimental evidence that all these various non-photosynthetic microbes contribute to the functioning of the whole lichen. Having said that, such experiments are also lacking for the photosynthetic partners too, although their utility appears so obvious that this is not usually seen as a problem. Even without such experiments, it looks very likely that some of these relationships aid the lichen as a whole – or aid the fungus, if you prefer a more fungal-centric view of these relationships – what is not clear is the extent of their importance. Are they a major part of a healthy lichen or just making a minor contribution to its success?

So what, then, is a lichen? It's clearly not a species, as it's composed of multiple species, but describing it as a mini-ecosystem doesn't seem right either – they seem more discrete, integrated and recognisably an entity than for example an oak woodland. That is why some naturalists specialise in lichens as a group; lichenologists exist because there is something superficially organism-like about lichens. Ideally we need a word that describes the concept of a composite organism in which a functioning entity is built from a number of closely integrated species. That word could be 'holobiont', but as with symbiosis the difficulty is that this is a term used in multiple ways by different scientists. As originally defined by Lynn Margulis (Fig. 53) in 1990, a holobiont was loosely defined as a composite organism comprising recognisable component parts – which certainly fits a lichen made of fungus along with photosynthetic and other microbes, and seems to describe a lichen's character better than 'mini-ecosystem'.

Later, however, Margulis and others, especially Eugene Rosenberg and Ilana Ziber-Rosenberg, started to use the term 'holobiont' in a more evolutionary context, suggesting that the holobiont itself was the target of natural selection and could evolve as an entity in much the same way that species evolve (O'Malley 2017, Lamm 2018). In general the term has been most widely applied to animals and their associated microbes – such as coral, which also has associated photosynthetic microbes, or insects and the host of microbes that live in their guts (Guerrero *et al.* 2013). With the rise of molecular methods this has become a large and fashionable area of research usually now referred to as microbiome research (the connections between the ideas of microbiome and Gaia are discussed at the end of Chapter 5). For example, there is now a substantial interest in the human microbiome, and the importance of the microbes living inside us for our health. These ideas can also be applied to plants, and there is now evidence that microbes can somehow move through the stolons connecting clonal plants – with such systems being referred to as meta-holobionts, since the holobiont is spread across multiple ramets (Vannier *et al.* 2019). I will return to these ideas later in the chapter after discussing some more examples of mutualisms – starting with mycorrhizae, which are also well represented in the Cairngorms.

MYCORRHIZAE

Two of the largest surviving tracts of Caledonian pine woods are in the Cairngorms National Park: Rothiemurchus, and just to the north the larger Abernethy Forest, which in autumn is a rich hunting ground for fungi. Amongst the pines, occasional birches and other broadleaved trees, fungal fruitbodies are scattered through the mossy ground layer. Some, such as the Fly Agaric (Fig. 105), are familiar from much of Britain, while many of the tooth fungi, such as Scaly Tooth (Fig. 106), are rarer specialists of the Scottish pine woods. Over 700 species of fungi have been found in the forest over the last 50 years or so, including 15 species of tooth fungi, one of which is probably new to science (Summers 2018, 2019).

FIG 105. Fly Agaric *Amanita muscaria* in Abernethy Forest. The photo gives a good representation of Abernethy, dominated by Scots Pine with an often moss-rich ground layer. However, Fly Agaric is a fungus most often associated with birch trees, and indeed there was a birch tree here too, just behind the photographer.

FIG 106. Scaly Tooth *Sarcodon squamosus* in Abernethy Forest: a mycorrhizal fungus associated with pines and in Britain mainly associated with Scottish pine woods. In older books this is often called *S. imbricatus*, a very similar-looking species which is now known not to be found in Britain (Kibby 2017).

The ecological roles of fungi are varied. As discussed in Chapter 3, many are important in breaking down and recycling organic matter, some can be parasitic (Fig. 107), while others form mycorrhizal mutualisms with many plant species. Mycorrhizae are a largely mutualistic symbiosis where fungi provide nutrients, and potentially other benefits such as protection from parasitic fungi, to a plant's root system. Some plants can be getting up to 80 per cent of their nitrogen and 90 per cent of their phosphorus via their mycorrhizae (van der Heijden & Horton 2009). In exchange the plant provides the fungus with the products of photosynthesis. Experiments show that in many (but not all) cases the plants grow better in the presence of their fungal partners, and something like 90 per cent of all plant species are involved in mycorrhizal associations. Indeed, in reviewing a century of research on mycorrhizal ecology, David Read (2002)

FIG 107. Rhizomorphs (root-like strands of fungal hyphae) of honey fungus *Armillaria* sp. growing on dead wood in Coed y Felin nature reserve near Mold in north Wales. There are a number of very similar-looking species of honey fungus in Britain. *A. mellea* and *A. ostoyae* are particularly virulent parasites infecting a large number of plant species, as well as being able to live on decaying dead wood (Spooner & Roberts 2005).

suggested that 'The major achievement of the first hundred years of research on the mycorrhizal symbiosis is the observation that the symbiosis is almost universally present in natural communities of terrestrial plants.' It follows from this that mycorrhizae really matter, being a major aspect of understanding the ecology of most plant species, although they often get at best only a passing mention in an introductory ecology course.

To get an idea of the potential importance of mycorrhizae, walk into a Bluebell wood in spring (Fig. 108). As Oliver Rackham (2003) enthused, these iconic Bluebell woods are 'one of the most glorious and distinctive types of British vegetation' – and without mycorrhizal fungi they wouldn't exist at all. Bluebells have very basic, poorly branched root systems and tend to grow on phosphorus-deficient soils, so they rely heavily on their fungal partners for this nutrient. Effectively much of what functions as a root system for these plants in fact consists of strands of fungal mycelia. These grow out of cells in the plant's roots and into the surrounding soil where they acquire nutrients, some of which are passed on to the Bluebell. So Rackham's 'glorious and distinctive' vegetation

FIG 108. Bluebells *Hyacinthoides non-scripta* flowering in the woodlands at the RSPB's Burton Mere reserve on the Wirral, Cheshire. Bluebells have very poorly developed root systems, so much of the nutrient acquisition from the soil is actually done by their mycorrhizal fungi, rather than directly by the plant. Technically these are arbuscular mycorrhizae, one of the two main types of mycorrhizal fungi (see Table 3). A study in North Yorkshire found at least six distinct types of mycorrhizal fungi associated with the Bluebell roots (Merryweather & Fitter 1995, 1998).

is mainly nourished by its mycorrhizae. It may be plants like Bluebell with poorly branched root systems that benefit more from mycorrhizal fungi for the uptake of soil nutrients such as phosphorus, while plants with more branched roots may mainly get other benefits, such as protection from other more parasitic fungi infecting their roots (Newsham *et al.* 1995).

Mycorrhizal fungi, however, are not all the same. Not only are they composed of many different fungal species, they can also be arranged into four main types. The two most important are the arbuscular mycorrhizae, which are found in around 74 per cent of plant species, and the ectomycorrhizae, which are associated with around 2 per cent of plants (Table 3). In addition, there are also more specialist mycorrhizae associated with orchids and heathers (van der Heijden *et al.* 2015). The mycorrhizae associated with Bluebells are arbuscular ones, mainly providing phosphorus to the plant, while ectomycorrhizae can sometimes also provide the plant with nitrogen from organic sources in the soil.

TABLE 3. Summary of key aspects of the two commonest types of mycorrhizal association – based on Figure 1 in Wilkinson (1998), heavily updated mainly from Table 1 in van der Heijden *et al.* (2015). There are two other less widespread types of mycorrhizae; orchid mycorrhiza and ericoid mycorrhiza, however we know much less about these compared to the arbuscular mycorrhiza (AM) and ectomycorrhiza. The main change between Wilkinson (1998) and this version is that the number of fungal species involved has increased. For AM the lower estimate of around 300 species is based on conventional microscopy of fungal spores while the higher value of 1,600 is based on estimates using molecular methods. In many cases there is little specificity, that is a particular fungal species can form mycorrhizal associations with many different plant species. Globally around 90 per cent of all plant species are involved in mycorrhizal associations (Spooner & Roberts 2005, van der Heijden *et al.* 2015).

Mycorrhizal type	Main plant groups involved	Fungal group involved	Does the fungus enter individual plant cells?	Estimated number of fungal species worldwide	Specificity of fungi
Arbuscular mycorrhiza (also called AM or V-A mycorrhiza)	Most herbs, grasses and many trees (especially tropical), also some hornworts and liverworts	Glomeromycota	Yes	300–1,600	None or very limited
Ectomycorrhiza	Conifers and some flowering plants (mainly temperate shrubs and trees)	Basidiomycota and Ascomycota	No	c.20,000	Variable, specificity in some cases

Although arbuscular mycorrhizae are the group associated with the largest number of plant species, it is the ectomycorrhizae that are the most apparent to a naturalist in the field. This is because the arbuscular mycorrhizae do not produce visible fruitbodies, while the ectomycorrhizae do, and many of our more recognisable mushrooms are mycorrhizal fungi (e.g. Figs. 105 & 106). In addition, most of Britain's major tree species have ectomycorrhizal fungal partners, so these fungi play a crucial role in many British habitats – indeed, this association of ectomycorrhizae with trees is found globally across the temperate zones. Worldwide, ectomycorrhizal fungi dominate forest where the rate of decay in the soil is limited by seasonal changes in temperature and/or moisture, an idea first discussed in detail by David Read in the early 1990s and now sometimes called 'Read's rule'. Recent estimates suggest that although ectomycorrhizae are only associated with around 2 per cent of all plants, approximately 60 per cent of all trees on the planet are ectomycorrhizal (Steidinger *et al.* 2019). These ectomycorrhizae get their name from the fact that the fungus sheathes the outside of plant roots (hence 'ecto'), while arbuscular mycorrhizae actually enter plant cells in the root, and are therefore sometimes referred to as endomycorrhizae (the term 'arbuscular' comes from the tree-like structures the fungus forms in the plant cell).

THE WOOD-WIDE WEB

The brief introduction to the natural history of mycorrhizae in the previous section describes a relatively simple situation where one, or a few, fungal species interact with an individual plant. While this may be true of experiments in plant pots in a university greenhouse, real life is more complex. Many mycorrhizal fungi show little specificity; that is, they can colonise a range of different plant species, rather than just specialising on a single plant. This is especially true for the arbuscular mycorrhizae, but is often the case with ectomycorrhizal fungi too (Table 3). For example, although most naturalists associate Fly Agaric with birches, it can sometimes be associated with other tree species such as Scots Pine or Beech (Spooner & Roberts 2005). What follows from this is important, because the same fungus can colonise not only multiple plants of the same species but also a range of different plant species; so different plants are linked underground by the fungal mycelia, forming a network. In the 1980s, when research on the ecological importance of such networks started to take off (e.g. Newman 1988), if anyone had been looking for a popular media-friendly metaphor to describe such networks I guess the one that would have come to mind was the telephone system, with many phones connected by a complex tangle of wires. A decade later, however, an obvious new alliterative metaphor had appeared, and the interlinking of plants by fungal networks started to be called the wood-wide web – a name that seems to have stuck.

The term 'wood-wide web' appears to have first been used on the cover of the journal *Nature* in August 1997 to describe research by Suzanne Simard and colleagues published in that issue – presumably coined by a member of *Nature*'s editorial staff with a felicitous turn of phrase. It was first used in the title of a research publication the following year, in a short report on the effects of agriculture on mycorrhizal diversity called 'Ploughing up the wood-wide web' (Helgason *et al.* 1998). This was also published in *Nature*, and Alastair Fitter, one of the authors, told me that they hoped that using this recently coined term in their title would attract the *Nature* editors to their paper and help get it accepted. It has more recently been further popularised by Peter Wohlleben in his widely read 2016 book *The Hidden Life of Trees*.

These mycorrhizal networks can potentially work rather like the plant stolons illustrated in Figure 99, moving resources between different plants – but in this case including plants of different species. One obvious possibility is that they could be moving carbon compounds, produced via photosynthesis, between plants. Laboratory experiments at the end of the 1960s appeared to show this happening, and such experiments were developed in much greater detail during the 1980s, especially by David Read and his colleagues at the University of Sheffield. The general approach was to use radioactive isotopes (Box 2) to 'label' carbon fed to one plant and then see if it moved into other plants linked to the original one via the mycorrhizal networks (Newman 1988). This can be hard to do under laboratory conditions but is even harder to do in the wild, especially with large plants such as trees. Away from the controlled conditions of the lab some studies have used stable (i.e. non-radioactive) isotopes of carbon to follow its movement between plants and fungi – as nitrogen was used in the study of nutrients in lichens mentioned above. The difficulties in conducting such

---

BOX 2. Isotopes – how to label your carbon

At the scale of description used in school science lessons, atoms are composed of a nucleus, comprising protons and neutrons, surrounded by a 'cloud' of electrons (these can all be broken down into more fundamental particles in more advanced physics). The protons and electrons carry a charge, and this affects how the atom reacts with other chemicals. The neutrons have no charge, and so just affect the mass of the atom (almost all the mass of an atom is in its protons and neutrons). An atom, such as carbon, can have a different number of neutrons but will still react chemically in the same way as other carbon atoms. However, because its mass is different it can behave differently in various physical processes (think of sorting items by weight). Atoms of the same element with different numbers of neutrons are called isotopes. Because different isotopes can be physically separated by their mass they can be used to 'label' molecules to track where they go.

Many isotopes are stable – such as carbon 12, which is by far the commonest isotope of carbon, and carbon 13. The numbers 12 and 13 are the mass numbers, indicating the number of protons plus neutrons in the atom. However, other isotopes are unstable and decay over time, and these are referred to as radioactive. One of the radioactive isotopes of carbon is carbon 14 (^{14}C), which has a half-life of 5,715 years. This means that it takes 5,715 years for half the atoms, on average, to decay, and the ratio of carbon 14 to stable carbon isotopes in a sample can be used to date archaeological samples (radiocarbon dating).

experiments partly explain why the importance of the wood-wide web to plant ecology is still controversial. However, it's clear that some plants at least can access quantities of carbon compounds from mycorrhizal networks, as they do so by parasitising these networks. A British example is the rather rare Yellow Bird's-nest *Hypopitys monotropa* – this has no green leaves and gets all its energy from the photosynthesis of other plants, accessing this by tapping into the mycorrhizae that are being fed carbon compounds from the surrounding trees.

An important early experiment addressing the effects of mycorrhizal networks on plant communities was carried out under laboratory conditions in Sheffield during the 1980s; it was led by Phil Grime (Fig. 19), with David Read providing the mycorrhizal expertise. They created plant communities in small pots (23 × 23 × 10 cm) of sterilised nutrient-poor lime-rich sand, comprising both grass and herbs from calcareous grassland. They then looked at the effects of both adding arbuscular mycorrhizae and disturbance by 'grazing' – the grazing being simulated by cutting the vegetation with scissors – over the course of a year. Both the 'grazing' and the mycorrhizae were associated with higher plant species diversity.[2] Why did adding mycorrhizae increase diversity? Grime and colleagues made a number of suggestions, firstly that some species such as Common Centaury *Centaurium erythraea* failed to get beyond seedling stage without mycorrhizae. However, the most important effect of the mycorrhizae was to favour the less dominant herbs over the dominant grass Sheep's-fescue *Festuca ovina*. Labelling of carbon compounds, from photosynthesis, with radioactive carbon appeared to show that some resources were moving from the most competitive plants to apparently subsidise the less competitive species (Grime *et al.* 1987, Grime 2001).

Can we interpret this as 'socialism in soil' – in the phrase coined by van der Heijden and Horton in a paper in the *Journal of Ecology* in 2009 for situations where the wood-wide web distributes 'wealth and power' more evenly? The answer appears to be a qualified 'yes' for at least some cases; however, this is still a controversial area. In their 'Socialism in soil?' paper van der Heijden and Horton reviewed the studies then available and concluded that in 45 per cent of the relevant published experiments seedlings benefited from access to mycorrhizal networks that were being maintained by other plants. The more difficult question is, why did they benefit? The most important reason is likely to be that the network provides a source of mycorrhizal mycelia that have already been supported, at no cost to the seedling, by the existing plants. However, as with Phil Grime's

---

2 The effects of grazing on plant diversity are discussed in more detail in Chapter 11, in the context of the intermediate disturbance hypothesis and other related ideas.

experiment, there is also evidence of the movement of carbon and other nutrients between plants – and it's this controversial idea that has attracted a lot of interest, in both the scientific and the more popular literature on the wood-wide web.

The controversy is not about the fact that some carbon can move through this web, but about how much, and whether it stays within the fungi in a plant's root or is able to enter the plant properly. The easiest way to understand the issue is to start by thinking of plants linked by stolons, where everyone is happy that resources are moving between the genetically identical ramets. Would it be reasonable to redraw Figure 99 to show mycorrhizal linkages between the plants, and plants of different species, rather than clones of the same species? Many of the key experiments have been done in North America, but one of the most interesting is from northern Switzerland, in a wood not too dissimilar to many you would find in Britain.

In 2009 I spent six months on a sabbatical research visit to the École polytechnique fédérale de Lausanne (EPFL) in Switzerland, and took the opportunity to take up an earlier invitation from Christian Körner, at the University of Basel, to visit one of his key research sites some 15 km south of the city. Driving out of Basel, we entered a largely agricultural landscape, the fields a patchwork of highly fertilised bright green plants, or vivid yellow expanses of oilseed rape flowers. On the hill slopes, above the flat agricultural land, are scattered woodlands with a mix of broadleaved and coniferous trees, and in one of these woods was Christian's research site. After driving down a track amongst the trees we parked in the shade and walked into a fenced-off part of the wood. The reason that this was a rather special research site was immediately apparent – for in amongst the trees was a large crane of the sort you usually see on a building site, but here used to access the woodland canopy (Fig. 109). Inside a gondola hanging from the crane you could be winched up from the dark of the woodland floor into the sunshine above, passing trees wrapped in pipes, perforated so that carbon dioxide could be pumped into the woodland canopy (Fig. 110). The idea was that this would allow carbon dioxide levels around the leaves to be raised to mimic a future in which humans have increased the concentration of the gas in the atmosphere, and to look at the effects on photosynthesis and water use by the trees. However, this system also allowed Körner and colleagues to use stable isotopes of carbon to follow the movement of carbon compounds from trees into mycorrhizal networks and potentially into surrounding plants too (Klein *et al.* 2016).

The experimental site was a mix of species, especially Norway Spruce, Beech, Scots Pine and European Larch. 'Labelled' carbon dioxide was released into the canopies of five spruce trees, and the research team found that about 4 per cent of these trees' carbon uptake was transferred into surrounding trees of the other three species, apparently via the ectomycorrhizal network in the soil.

ABOVE: **FIG 109.** The Swiss canopy crane, photographed in 2009. The crane was 45 m tall with a 30 m jib providing access from a gondola to 64 trees over approximately 3,000 m² of woodland (Leuzinger & Körner 2007). The woodland is a mix of broadleaved and coniferous trees – the species involved in the wood-wide web study were Norway Spruce *Picea abies*, Beech *Fagus sylvatica*, Scots Pine *Pinus sylvestris* and European Larch *Larix decidua*. The crane has now been dismantled following Christian Körner's retirement.

FIG 110. Piping twisted through the foliage of a spruce. This was used to release labelled carbon dioxide into the tree's canopy, which the tree then used in photosynthesis. Photographed in 2009 from the gondola suspended from the canopy crane.

Clearly finding this carbon in nearby trees doesn't by itself show that it had moved via the wood-wide web; it could for example diffuse through the air or be released into the soil and then taken up by the roots of nearby trees. To try and rule out these other possibilities this study also looked for labelled carbon in the fruitbodies of the ectomycorrhizal fungi forming a network with the trees (and found it). They also looked in the fruitbodies of non-mycorrhizal fungi (and didn't find it), and in woodland herbs which are involved in arbuscular mycorrhizal networks, rather than the trees' ectomycorrhizal ones (and didn't find it). So the labelled carbon was only turning up in plants and fungi attached to the ectomycorrhizal network.

Although doing a field experiment at this large woodland scale, with 40 m trees, is difficult, and it's hard to be sure you have covered every possibility, it looks very likely that relatively large amounts of carbon were moving through the wood-wide web into adjacent trees. Indeed, this study is particularly important as it shows larger amounts of carbon moving in this way than other similar experiments – though few other experiments have been able to use such tall trees. This Swiss study provides some of the best evidence so far for substantial movements of carbon between trees via the wood-wide web.

Over the last decade or so a small number of studies have started to suggest other roles for mycorrhizal networks – in transmitting messages between plants. Experiments, mainly so far with crop plants, have demonstrated the movement of signals warning of pathogen or aphid attack between plants via the network. This work is all rather new, little has been done on plants growing in the wild, and it's not yet clear how important these processes are. However, it has the potential to be important for both plant ecology and agriculture in the future (Babikova *et al.* 2013).

MIXOTROPHY AND SYNTROPHY – ECOLOGY'S MISSING TERMS

Lichens and ectomycorrhizae are the two symbiotic mutualisms most easily visible to a naturalist in the field, but if we turn to the microscopic world we will find that mutualisms are very extensive indeed. As described in Chapter 1, a seemingly fundamental distinction made in ecology is to divide organisms into autotrophs and heterotrophs. Plants are the classic example of autotrophs, but a few moments' thought starts to come up with some exceptions. Think of carnivorous plants such as sundews *Drosera* spp., or a plant such as Mistletoe (Fig. 111), which is simultaneously photosynthetic and parasitic. Carnivorous plants are mainly using their animal prey as a source of nutrients, such as nitrogen or phosphorus, but may in some cases also get carbon compounds from their prey too. While mistletoes can make carbon compounds by photosynthesis in their own leaves, they may get over half of the carbon they need from photosynthesis by their host tree. Such plants are mixotrophs – that is, they live as both autotrophs and heterotrophs, getting some of their carbon from sources other than their own photosynthesis. However, other less obvious plants may be mixotrophic too. Consider the trees in the experiments of Körner and colleagues in Switzerland – these were apparently getting some of their carbon from other trees via the mycorrhizal network in the woodland soil. The Common Twayblade (Fig. 3) is another interesting example. This orchid looks like a straightforward autotroph, as it has broad green leaves and can obviously photosynthesise. However, it is also receiving useful amounts of carbon from other plants via its mycorrhizae (Selosse & Roy 2009).

Although we are now finding mixotrophy to be much commoner than previously thought amongst plants, it's really in the microbial world that it

FIG 111. Mistletoe *Viscum album* near Hay-on-Wye on the Welsh/English border. It is a mixotroph, obtaining its carbon both from its own photosynthesis and from the tree it parasitises.

becomes a major way of life – breaking down our intuitive feeling that organisms *either* photosynthesise *or* eat other organisms. Many microbes do both, although you will struggle to find the term 'mixotroph' in many ecology texts, despite the fact that this way of life is common in many eukaryotic microbes living in habitats that give them access to light (Schmidt *et al.* 2013). Many of these are planktonic, living in freshwater or marine conditions, but mixotrophic microbes can be found in other places too.

Mixotrophy is thought to be a particularly advantageous strategy in nutrient-poor environments such as the ombrotrophic bogs – where the only input of nutrients comes from the few that happen to be dissolved in rain water. Look into the tiny ecosystem to be found in bog moss at such sites with the aid of a microscope, and protists harbouring photosynthetic algae can be found (Fig. 112). Clearly these protists benefit by keeping the algae alive and allowing them to photosynthesise, and in addition they may also sometimes consume some of their stock of algae. But do the algae benefit in any way? The simple answer is that we are not yet sure, although the big advances that molecular methods are making in microbial ecology mean that we may soon start to have clearer answers. The likely origin of most of these algae is that they have been taken in from the environment – although plausibly some may be inherited from a parent. So are they effectively just 'food' for the protist? Probably not, as there is some suggestion that these algae have adaptations for life within the protist (Lara & Gomaa 2017) – and if so this suggests this is a mutualism, with both partners benefiting from the interaction. So mixotrophy may cover a spectrum from predation (some carnivorous plants), through parasitism (Mistletoe), to mutualism in the case of some eukaryotic microbes. Certainly, mixotrophy itself is much more widespread than commonly assumed.

Mutualisms in the microbial realm go far beyond the possibilities associated with mixotrophy, and are often described by the technical term syntrophy. The basic idea is familiar from human economics, and the opening of a review paper by Morris and colleagues (2013) makes the point nicely:

> When complex, difficult jobs have to be done, it is wise to divide the work into smaller, simpler tasks and to include specialists to ensure a positive and worthwhile outcome for all team members. This principle of labour division is also true for microorganisms.

The idea is that two or more species of microbes working closely together can often break down and use chemicals that neither of them could utilise by themselves. These mutualistic symbioses are known to be widespread –

especially, but not exclusively, in oxygen-poor environments. Some organisms that have in the past been thought to be a single microbial species are now known to be a holobiont comprising two cooperating microbes, neither of which could survive without the other. A well-known example is '*Thiodendron latens*',

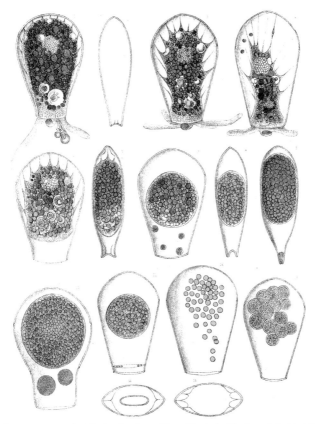

FIG 112. The testate amoeba *Hyalosphenia papilio* as illustrated by Joseph Leidy (1879). In the days before good photographs could be taken using microscopes it really helped to be an accomplished illustrator if studying microbes (indeed, it's still a useful skill, even today), and Leidy was a master of the art. The green spheres are all algal cells living within the protist and providing carbon compounds from photosynthesis for the protist to use. This amoeba is a characteristic microbe of habitats rich in bog moss (*Sphagnum*), such as those studied by Bill Heal at Moor House (discussed in Chapter 4 – see Fig. 61). Heal found this species in his studies at Moor House, although recent molecular studies suggest that there are several distinct genetic lineages of the species, which are impossible to tell apart using microscopy but show differences in the climatic conditions in which they live (Heger *et al.* 2013).

first described as a new bacterium from Russia in the 1960s and later found in a range of aquatic sites around the world. However, what was first thought to be a single species, albeit with a complex-looking life cycle, has turned out to be two different organisms working together to make a living breaking down sulphur compounds (Margulis *et al.* 2000).

A more spectacular example is the role of syntrophy in the production of methane by trees. On occasion trees can produce substantial amounts of methane gas, to the extent that a corer driven into the heartwood of a tree may release quantities of flammable gas that can be ignited, burning with the bluish flame you associate with a gas hob in the kitchen. There are multiple sources for this methane – including from the soil and from the non-biological breakdown of some plant chemicals. However, some of it comes from microbes – archaea – which break down plant chemicals such as cellulose and give off methane as a waste product. This is a complex process, and the archaea appear to need the help of a consortia of bacteria and fungi to create the conditions which allow them to do this (Covey & Megonigal 2019).

So syntrophy is very common in microbes, especially prokaryotes, and raises interesting evolutionary questions which are briefly discussed in the next section. It also raises an interesting question about the emphasis given to competition in mainstream academic ecology. If ecological theory had developed with a greater input from microbial examples, would we have given competition quite as much emphasis? Clearly microbes compete, but mutualistic interactions seem to be much commoner than in the macroscopic animals that have dominated much of ecological and evolutionary theory in the past.

EVOLUTIONARY QUESTIONS

Cooperative symbiotic mutualisms, such as those discussed in this chapter, create an apparent problem for theoretically inclined naturalists – namely, how can they evolve? At its simplest the question is this – why don't the partners in these interactions cheat, by taking the benefits without providing anything in return? The obvious analogy is with the tax system in a welfare state. The temptation (without the intervention of the tax authorities) is for people to take advantage of the benefits but to skip paying the taxes which underpin those societal benefits. Why doesn't this happen in mutualisms?

Some answers to this question are well established, such as the idea that so-called 'vertical transmission' aids cooperation. Vertical transmission is where both partners in the relationship are dispersed together. For example, some grasses contain fungi which help protect them from grazing, and these fungi can be found in the grass seed. So the fates of both partners are closely tied together at reproduction, making cooperation rather than cheating important for their joint futures. Lichens can sometime reproduce via fragments breaking off the lichen's body – again in this case all partners are transmitted 'vertically' into the next generation. However, lichens can also reproduce by spores, which may only contain the fungal partner, and this has to be re-infected from the environment by new algae and other microbes – so-called horizontal transmission. So although vertical transmission seems likely to facilitate mutualisms, many ecologically important mutualisms (such as those given the most attention in this chapter – namely plants and mycorrhizal fungi, and many cases of lichens) are horizontally transmitted, re-forming each generation. Here 'cheating' would seem more likely, because successful cooperation is not as closely tied to reproduction.

So how can horizontally transmitted mutualisms be stable and avoid breaking down under the influence of cheats? A number of possibilities have been suggested. As in population ecology (Chapter 8), spatial structure may be important. In the case of mycorrhizae, many plant seeds don't travel very far and so land on soil close to their parent and therefore have a good chance of forming relationships with the same fungi that colonise their parent's roots. This is not technically vertical transmission, as the fungus is not transmitted in the seed, but it can have the same effect, and is sometimes called pseudo-vertical transmission.

Other possibilities include some sort of policing – as with human taxation – with sanctions being taken against a partner that doesn't cooperate. It may also be the case that there are counteracting advantages to horizontal transmission, such as the ability to pick a partner suitable for the particular local conditions rather than being stuck with the one you inherited from a parent – an idea often referred to as 'biological markets'. In many cases of syntrophy, cheating may be difficult if the two organisms can only utilise a food source by working together.

All these ideas have been around for some time – indeed, the preceding three paragraphs effectively summarise a technical article on this topic I co-wrote some two decades ago (Wilkinson & Sherratt 2001). However, more recently other ideas have been suggested. For example, Ford Doolittle's ideas on selection by survival (described in Chapter 5) look relevant to holobionts such as lichens. As I have already pointed out, he has told me that it was dissatisfaction with the way some scientists were suggesting that communities of microbes – for example the human microbiome – could be the subject of natural selection that triggered his thinking in this area. There are a host of potential explanations for how mutualisms evolve and maintain stability. The one thing that is clear is that there is no single explanation – one size (or theory) doesn't fit all.

CONCLUDING REMARKS

The impression you get from both TV natural history programmes and ecology textbooks is that competition is *the* driving process of what we see in nature. The same seems to follow from evolution, often characterised as 'survival of the fittest', with competition between organisms often being a key decider of 'fitness'. Competition is undoubtedly important (see Chapters 6 and 10), but evolution can also lead to apparent cooperation, and such mutualisms appear to be common – especially amongst microbes.

This chapter has concentrated on symbiotic mutualisms, those mutually beneficial interactions where the individual organisms can be hard to separate, such as lichens, or mycorrhizal fungi in plant roots. These two examples are described in some detail, because lichens and ectomycorrhizal fungi are particularly obvious to a field naturalist, and if this were a book taking its key examples from gardening the nitrogen fixation in legumes would likely have been a central example. However, as the discussion of mixotrophy and syntrophy makes clear, mutualisms appear especially common in the microbial realm. Overall, symbiotic mutualisms are globally important processes, as indeed are non-symbiotic ones such as many examples of insect pollination and seed dispersal by animals. In fact, all of the topics discussed in this chapter are more significant in ecology than they might at first sight appear: syntrophy is important in breaking down chemicals in many nutrient cycles, lichens dominate the vegetation in many high-latitude and high-altitude environments, and mycorrhizae are key to the success of most of the world's terrestrial plant species. The recent advances in applying the techniques of molecular biology to microbial ecology mean that this is an area ripe for exciting advances over the next decade or two.

CHAPTER 8

Can We Explain Selborne's Swifts?

I t is early summer, and after a cold late spring the weather is now very warm. In the small churchyard the shade from the yew tree is welcome, and from this patch of relative coolness I watch swallows, martins and swifts hunting in the insect-rich air. Away from the village the parish is an intimate mix of woodlands, heath and agricultural fields – with the dung from the local cattle providing good breeding grounds for flies and other insects, which then feed the birds swooping round the church tower (Fig. 113). Attempting to count the fast-moving birds against the clear blue sky, it's clear that the

FIG 113. View of Selborne Church from the south, from an 1875 edition of *The Natural History of Selborne*, clearly based on a drawing by Hieronymus Grimm commissioned by Gilbert White for the first edition.

FIG 114. The Wakes, Gilbert White's house, now a museum. Photographed from the wider parkland looking into the garden over the ha-ha – designed to keep livestock out of the garden without obscuring the view. This ha-ha, added by White, was one of the first to be built in Britain in a smaller private garden, rather than at a large stately home (Mabey 1986).

swifts are less common than the swallows and martins. Being fewer in number makes the swifts easier to count, and there seem to be eight pairs, with half of these nesting in the church in front of me, while a short walk round the village suggests the rest are nesting in the thatched roofs of local cottages.[3]

This opening paragraph is a fiction. It depicts the Hampshire village of Selborne in the 1770s as described in Gilbert White's classic book of 1789, *The Natural History of Selborne*. Today the village is still off the main roads in a now crowded southern Britain, but it's nowhere near as isolated as it was in the eighteenth century. The small, and now sometimes congested, village car park is surrounded by trees providing shade on a hot day like the one I imagined from 1778, and from here it's a short walk down the main street to The Wakes – Gilbert White's house, which is now a museum devoted to the life of White and also Captain Oates of Antarctic fame. Beyond the formal garden behind The Wakes lie hay meadows managed for their wildlife, and much of this less formal 'Park' was cut for hay in White's day too (Figs. 114 & 115). When I visited in August 2015, as the meadows were being cut, they were full of the dried

---

3 The historical reconstruction that opens this chapter was guided by Gilbert White's classic book *The Natural History of Selborne*, along with information from Richard Mabey's 1986 biography of White, the illustration of Selborne Church shown in Figure 113, and Kington's 2010 New Naturalist volume on the history of Britain's climate and weather.

FIG 115. Haymaking at The Wakes, summer 2015.

flowers of Yellow-rattle *Rhiananthus minor*, living up to their name with the
seeds rattling inside, while grasshoppers called in the drying hay. Cut out the
background noise of the twenty-first century and focus your gaze carefully to
exclude anything obviously modern, and you could be back in the eighteenth
century – and in a countryside much more diverse in its wildlife than the wider
landscape is today.

White's famous book is constructed as a series of letters to two
correspondents with natural history interests (some of the letters were actually
sent, while others were composed especially for the book). It describes the
natural history around Selborne, and according to a leading historian of ecology
it 'achieved a level of precise observations and depth of reasoning that set a new
standard for natural history' (Egerton 2012). One of the birds that interested
White was the Swift (Fig. 116), and this is what he wrote about the population in
Selborne (in Letter 39 to Daines Barrington, dated 13 May 1778):

> *Among the many singularities attending those amusing birds the swifts, I am*
> *now confirmed in the opinion that we have every year the same number of pairs*
> *invariably; at least the result of my inquiry has been exactly the same for a*
> *long time past … The number that I constantly find are eight pairs; about half*
> *of which reside in the church, the rest build in some of the lowest and meanest*
> *thatched cottages. Now as these eight pairs, allowances being made for accidents,*
> *breed yearly eight pairs more, what becomes annually of this increase; and what*
> *determines every spring which pairs shall visit us, and reoccupy their old haunts?*

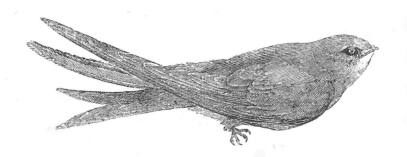

FIG 116. A Swift *Apus apus* from Thomas Bewick's *History of British Birds*, first published at the end of the eighteenth century (this illustration comes from an 1821 edition). Bewick used White's observations in his book, especially in his list of dates of arrival in Britain of migratory birds – in the case of the Swift he cites late April. In addition, Bewick's illustrations have been used in multiple editions of *The Natural History of Selborne*.

White's account is unusual for the time, both in presenting semi-quantitative population data – around eight pairs over many years – and in asking *why* this should be the case. This chapter is all about modern attempts to answer the 'why' question about population sizes.

Move forward in time to the early 1980s, and two eminent ecologists in their forties, struck by Gilbert White's description of eighteenth-century Swift populations, decide to go to Selborne to have a look. The two were John Lawton (a keen birdwatcher) and Robert May. John remembers telling Bob May about Gilbert White's observations on Swift populations in Selborne during a walking holiday in northern England. May was intrigued, and suggested they had a trip to Selborne to see if the numbers were still similar to White's count of eight pairs. When they visited they found around 12 pairs of Swifts nesting in the village, which they felt was amazingly close to the eighteenth-century value of eight – so the Selborne Swift population hadn't changed much in the intervening two centuries, despite many changes in the village and surrounding countryside over that time. May suggested they write an essay for the journal *Nature* using the Selborne Swifts as an intriguing historical example illustrating the kind of questions population ecology tries to answer (Lawton & May 1983).

The apparent stability of the Selborne Swift population so impressed May that several decades later he also used it as the opening example in the third edition of his book on theoretical ecology (McLean & May 2007), and John Lawton also returned to this example in later writing (Lawton 1999b). Since the 1980s Swift populations in Britain have tended to decline – but in Selborne in 2018 there were still around a dozen nesting pairs, although additional nest boxes were added early in 2019 so possibly numbers will increase in future (Norris 2020). How can we explain this stability, or indeed the lack of stability seen in the population of British Swifts as a whole? A start on understanding the basic ideas that may explain how and why populations vary over time can be found from looking at a pond covered with duckweed. However, before considering duckweed, it's worth briefly thinking about what we mean by a 'population' in ecology.

WHAT IS A POPULATION?

In ecology a 'population' is any group of organisms of the same species living at a particular location in time and space. Consider the large common ground beetle *Carabus problematicus* (Fig. 117). Many species of ground beetles have been

FIG 117. The ground beetle *Carabus problematicus* – several similar-looking large *Carabus* species are found in Britain. Analysis of data from across the UK Environmental Change Network sites shows that many species of ground beetle are declining, and also that many are starting their activity earlier in the year as the climate changes. *C. problematicus* has shown limited changes, with only a modest decline in its abundance and no significant change to its activity early in the year. The greatest declines in ground beetles have been detected at sites in the mountains (Brooks *et al.* 2012, Pozsgai *et al.* 2018). The individual illustrated was photographed just outside the boundary of the Environmental Change Network site on Snowdon, in the mountains of north Wales.

declining over the last quarter of a century, making their population ecology of some conservation interest. For a beetle such as *C. problematicus* one could ask about the size of its global population (it's widely distributed across Europe), or the UK population as a whole, or the size of a population at a specific site. At the scale of a particular site the normal approach for large active ground beetles is to use pitfall traps to estimate numbers. These traps are containers sunk into the ground that the beetles fall into as they run over the ground surface. In the UK, data of this type have been collected from 12 Environmental Change Network sites scattered across the country where lines of such traps are monitored on a regular basis, with pitfall data going back to 1994 or 1999 depending on site (Brooks *et al.* 2012). Indeed many of the classic sites used to open chapters in this book are part of the Environmental Change Network, including the Cairngorms, Moor House, Rothamsted and Wytham Woods. This network of sites provides long-term information on changes to the UK's environment. With data from these sites, covering a range of different locations and habitats, mathematical approaches can then be used to attempt to summarise ground beetle trends at the UK level.

This highlights an important aspect of population ecology. Because its key questions are about numbers – and changes in numbers over time – then it's relatively easy to see how to apply mathematics to these questions. Usually in science, if maths can usefully be applied to a problem this often helps in getting at an answer. However, in this chapter I will describe some of the key ideas of population ecology without mathematical detail, illustrating the ideas with simple graphs and diagrams.

The ground beetle example also helps illustrate another point. Many of the examples in this chapter are of animal populations, and there are good reasons for this as usually with animals it's quite obvious what an individual is – e.g. a single beetle that has fallen into your pitfall trap. With plants, fungi and some other groups this can be much less obvious. Think of the vegetation in a meadow and then ask yourself how many individuals of a particular species of grass are present? Or how many individuals of White Clover – which can be a vertically spreading group of genetically identical plants rather than discrete countable individuals. Therefore, to keep things simple while introducing the basic ideas, much of this chapter uses animal examples. But let's first consider an instructive plant example.

WHAT CAN DUCKWEED TELL US?

To get a feel for several key ideas in population ecology, all you need do is consider a pond or drainage ditch covered with duckweed (Fig. 118). There are

FIG 118. Duck and duckweed: a drake Mallard *Anas platyrhynchos* swimming through duckweed at the Wildfowl & Wetlands Trust Martin Mere reserve. Duckweed is indeed food for ducks, and in addition ducks play a role in moving the plant between ponds. As can be seen in the photograph, the plants can stick to the bird's feathers and bill, and in addition it has recently been shown in South America that duckweed can pass through the digestive system of ducks and swans and still be viable (Silva *et al.* 2018). This presumably happens with British duckweed species too, but has not yet been demonstrated.

several species of these tiny flowering plants found in Britain (some introduced), with Common Duckweed *Lemna minor* being the one most commonly seen. The duckweeds, which often form a green covering over small bodies of still water, are an interesting group, and include the world's smallest flowering plants (the tropical and warm temperate *Wolffia* species, one of which, Rootless Duckweed *W. arrhiza*, is naturalised in southern England). Much of their reproduction is by asexual budding, where one plant gives rise to a new individual. The population can therefore double each generation, and this lends itself to an instructive simple maths problem.

For simplicity, assume that no duckweed dies, or is eaten by ducks, and let's also assume that the amount of duckweed doubles every day. If it takes 48 days for the patch of duckweed to cover the whole pond, then how long would it take to cover half the pond? This question (using the less botanically realistic waterlilies rather than duck weed) was used as one of the questions in a survey of basic scientific and mathematical literacy of 2,000 people in the USA by Dan Kahan (2017). Only 23 per cent got the correct answer of 47 days. The 77 per cent who got it wrong are not alone in underestimating this type of population 'doubling' –

exponential growth being the correct mathematical term for this type of growth. The common illustration used in many maths texts is the story of a (presumably apocryphal) king who offered a reward to one of his favoured subjects. The reward asked for was a grain of wheat on the first square of a chess board, two grains on the second, four on the third, and so on until the final square. This sounded trivial to the king, who readily agreed. However, you end up with enough wheat to make a pile around the size of Mount Everest – that is, just under 2^{64} grains (Wilson 2018). The power of exponential growth to generate large numbers very quickly has come to wider attention with the COVID-19 pandemic. Governments which were slow to impose lockdowns quickly found that exponentially growing numbers of cases became a major public health disaster. Exponential growth can be a matter of life or death, and the basic idea is shown graphically in Figure 119.

Exponential growth is important because it's not just duckweed that looks as if it should grow like this; much of biological reproduction has this doubling character – albeit in a slightly more complex fashion for sexually reproducing organisms. Certainly any population seems to have the ability to produce enough offspring to create a rapidly expanding population. This was Gilbert White's point about Swifts – if they are producing all those offspring, why does the

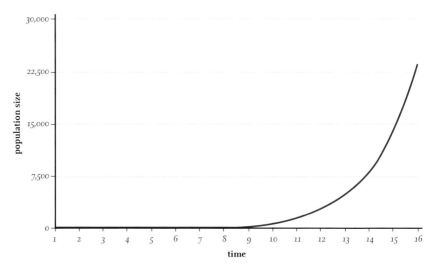

FIG 119. Exponential (also called geometric) growth – think of filling the first two rows of a chess board with wheat grains as described in the text, or duckweed expanding across a pond surface. To start with, the rate of growth looks slow, but with time it appears to really take off. In fact the growth rate doesn't change (in this case it's always 'multiply by two' each time step) but the effect is a rapidly increasing population.

population stay the same from year to year? As John Lawton (1999b) pointed out in an essay on White, the really clever thing is that he had noticed the importance of something that doesn't happen (ever-increasing numbers of Swifts) despite simple mathematical reasoning suggesting it should happen. Clearly exponential growth is not the full story; other things must be going on.

As biological populations have this potential for very rapid growth, something must intervene to restrict population size and prevent unrestrained exponential growth continuing for prolonged periods. The importance of this was stressed by Thomas Malthus in his famous *Essay on the Principle of Population*, which was published a decade after *The Natural History of Selborne*. Today this book is perhaps best known to naturalists because of its influence on Charles Darwin's ideas on natural selection. Malthus pointed out that human populations potentially increase exponentially, while our ability to grow food increases more slowly – inevitably leading to population control by starvation or other disasters if human population growth remains unchecked (Egerton 2012).

As long-term exponential growth seems an unlikely scenario, a more realistic description of population growth would be a population that grows quickly to start with but levels off rather than expanding forever. One of the simplest mathematical equations you can construct with these characteristics is the logistic equation, which is shown graphically in Figure 120. Here the population levels off at a value referred to as the carrying capacity (in textbooks usually referred to by the symbol K), the idea being that this is the maximum number of individuals that the environment can support. The basic idea of carrying capacity is nicely illustrated by duckweed or other floating water plants: once a pond surface is fully covered in the plants then the population cannot grow any more as it has run out of space (Fig. 121). If your favourite habitat is a rocky shore rather than a freshwater pond, then think in terms of barnacles covering a rock.

The idea that you could describe the growth of a typical population using the logistic equation was conceived multiple times in the late nineteenth and early twentieth centuries. It is most often associated with Raymond Pearl (working with the mathematician Lowell J. Reed in 1920); however, unknown to Pearl at the time, it had been suggested by Pierre-François Verhulst some 80 years before, and indeed the term 'logistic' is Verhulst's. From the 1920s onward it became a focus of controversy. Was it an accurate mathematical description of population growth (Pearl liked to refer to it as a scientific 'law'), was it a good enough approximation to be useful, or was it just plain wrong (Kingsland 1995)?

The great strength, and also the weakness, of the logistic description of a population is its simplicity – it's one of the simplest mathematical descriptions of population growth that combines exponential growth at small population

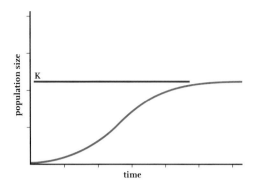

FIG 120. A graphical representation of logistic growth. At low population sizes the growth is exponential, but growth slows as the population becomes larger. At the carrying capacity (K), growth is zero. The logistic equation which generates this graph aims to describe a population with overlapping generations; a similar equation describes populations with non-overlapping generations (e.g. dragonflies, where one year's adults die off before the new adults emerge the next year). More technically (for readers with a background in calculus) the logistic equation is a differential equation, while the one describing non-overlapping generations is a difference equation. Mathematically the important difference is continuous reproduction (hence calculus) as opposed to discrete reproductive events (hence recursion).

size with the realistic idea that a population will level off at some maximum value. To do this it multiplies together just three factors to predict the future population size: the current population size, a constant growth factor (usually referred to as r), and the size difference between the carrying capacity and the current population size, all divided by the carrying capacity. This simplicity makes it a good starting point for thinking about populations, but also means it's unlikely to often exactly describe the reality of any particular case. Consider the idea of carrying capacity. There clearly has to be a limit to the numbers of Swifts that can live in Selborne, but it's also likely that this limit varies over time – there is for example not much Swift food available in Hampshire in winter, which is why Swifts migrate south after the breeding season, and the amount available in early summer will vary from year to year (and from day to day). So K in Figure 120 will not be an unvarying number, except perhaps in simple cases like a pond covered with duckweed (always assuming that the area of water in the pond doesn't change).

At least in some illustrative cases, where a simplified system has been created in the laboratory, something like the logistic growth curve can be seen in real data. A classic example – still described in modern textbooks – is the work of

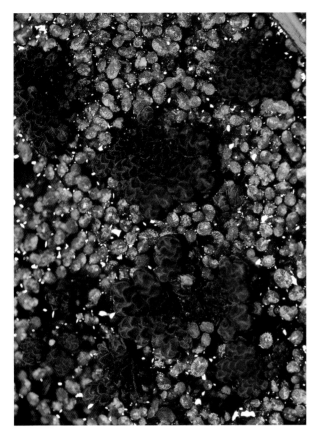

FIG 121. The idea of carrying capacity is nicely illustrated by floating aquatic plants. Once the entire surface of the pond is covered, then the population will stop growing. Duckweed and Water Fern *Azolla filiculoides* on a pond in the Isles of Scilly. The fact that there is more than one plant species shown in the photograph suggests that interspecific competition will also be involved in this example (as described in Chapter 6).

the Russian Georgii Gause in the 1930s. He used simple lab systems with ciliate protists (several species of *Paramecium*) feeding on microscopic yeast cells. His data gave graphs of a similar shape to Figure 120, and by altering the amount of food (yeast cells) he was able to increase or decrease the carrying capacity of his lab system for *Paramecium*. This was a clear demonstration of the idea of logistic growth and carrying capacity – hence its regular appearance in textbooks. However, we now know it's actually a nice example of a scientist getting the 'right' answer for the 'wrong' reasons. While various species of *Paramecium* can indeed ingest yeast – you can watch them doing so under a microscope – they can't successfully grow on such a diet (Pritchard *et al.* 2016). What appears to have happened is that Gause's yeast cultures were contaminated by bacteria which are good food for *Paramecium* (it's still not straightforward to avoid contamination like this, and it would have been even harder with 1930s lab technology).

In concentrating the yeast to give the *Paramecium* more food, he presumably also concentrated the bacteria – the actual food source. So the cultures with high numbers of yeast cells did have more protozoan food in them, it just wasn't the yeast that Gause thought was the food. I will return to these classic experiments in the chapter on ecological niches (Chapter 10).

To get a feel for the difficulties for real-world (rather than laboratory) populations, consider the animal we have studied in most detail – ourselves. There have been multiple attempts over the last few centuries to calculate the carrying capacity of Earth for humans (Cohen 1995). In 1679 van Leeuwenhoek came up with a figure of 13.4 billion (remember a billion is a thousand million) by assuming that the inhabited parts of the Earth could potentially support a population at the density of people found in the Netherlands at the time. The first estimate using the logistic equation was by Pearl and Reed in 1924 – they used estimates of global population between 1660 and 1914 to predict a carrying capacity of 2 billion, assuming that global population was following a logistic curve. Around a decade later Pearl had increased this estimate to 2.6 billion. Estimates by other ecologists mentioned in this book include 5–7 billion by Robert Whittaker and Gene Likens in 1975, where they assumed that everyone would have to be supported by peasant agriculture as they were unconvinced about the long-term future of intensive agriculture. The complete opposite of this was the astoundingly optimistic figure of 300 billion proposed by Garrett Hardin in 1986, where he assumed US-style energy use across the whole planet supporting intensive agriculture, although he admitted this was 'only a crude estimate'. To give these figures some perspective, the best estimate of the number of humans on the planet at the end of 2020 was almost 7.9 billion.

What does this wide range of estimates tell us? First, it's obviously hard to calculate the human carrying capacity of the planet! Second, as technology changes so does the potential carrying capacity, so Pearl's estimates from the 1920s based on fitting a logistic curve to data from the preceding 250 years couldn't factor in changes in agriculture during the twentieth century. Something similar will apply to other organisms, as it's difficult to guess how important aspects of their environment will change with time. In addition, when thinking about humans, a further difficulty in defining what we mean by carrying capacity becomes clear – is this the carrying capacity for a planet of subsistence farmers, or something allowing a more attractive lifestyle? Joel Cohen (1995) ended his important book on human carrying capacity with an extended quotation from the nineteenth-century philosopher John Stuart Mill, highlighting the fact that we might want a population substantially less than the theoretical maximum that could find enough food to eat. Mill wrote:

Nor is there much satisfaction in contemplating the world with nothing left to the spontaneous activity of nature, with every rood of land brought into cultivation, which is capable of growing food for human beings; every flowery waste or natural pasture ploughed up, all quadrupeds or birds which are not domesticated for man's use exterminated.

I suspect most readers of this book will be sympathetic to Mill's view of the importance of the natural world to human quality of life.

Returning to considering non-human organisms, if we accept the basic truism that populations cannot grow indefinitely and that some factors must act to counteract this growth, then the question is, what are those factors? The logistic equation doesn't answer this question, it just assumes that population density is in some way important in the process. How realistic is this, and what factors might be involved? A good starting point is to consider the Grey Heron population of Britain (Fig. 122).

FIG 122. The Grey Heron *Ardea cinerea* is widely distributed across Eurasia and Africa. It's found in a wide range of freshwater habitats, where fish are usually its main prey. The population in Britain is the subject of the longest-running population study of any bird species.

WHAT CAN HERONS TELL US?

To understand population ecology it obviously helps to have runs of data over multiple years, and birds are particularly well studied from this point of view. Indeed there is a whole book in the New Naturalist series on British bird populations that's almost 600 pages long (Newton 2013). Amongst all these data, those on the Grey Heron in England and Wales stand out as of truly international significance, as they are the longest-running survey of the numbers of a breeding bird anywhere in the world. Breeding numbers of Grey Herons are unusually easy to survey as they build large nests, often high in trees, and occupy them early in the year before full leaf cover has developed (Fig. 123). Taking advantage of this, and the large number of birdwatchers in Britain, Max Nicholson started a national census of herons in England and Wales in 1928 using volunteer observers (those observers were then too thin on the ground in Scotland and Northern Ireland for the approach to work there). This study continues to this day and is now organised by the British Trust for Ornithology (BTO) – indeed, Max Nicholson was one of the founders of the BTO only a few years after the

FIG 123. In late February 2019 the heronry at the Yorkshire Sculpture Park was already showing the first signs of occupation by breeding birds. Occupied nests are easy to count, and the long-running census of breeding herons in England and Wales counts occupied nests in the second part of April, ideally supplemented with data from March and May too.

heronry census started. These data have been supplemented by occasional more detailed surveys of heron numbers, which can be used to help control for factors such as variations in the numbers of volunteer nest counters from year to year (Marchant *et al.* 2004).

Ninety years of Grey Heron population data are shown in Figure 124. Two things stand out when looking at this graph: first, there have been some substantial ups and downs in the population over the last century, and second, despite these occasional crashes the number of breeding herons appears to fluctuate around a long-term average value – with possibly a slight tendency to increase over the years of the survey. Consider the large fluctuations first. The most dramatic of these is in the early 1960s, and in fact this crash followed the winter of 1962/63. This was the coldest winter since 1740, and in many places in Britain there was continual snow cover for 10 weeks from late December until early March, with substantial snowdrifts even in London and ice floes in the Thames (Kington 2010). Indeed, early in 1963 ice floes were common around much of the British coastline (Fig. 125). For a bird that largely feeds on fish, such extensive cold weather, and associated frozen water bodies, is very bad news, hence the spectacular decline in heron numbers. The other main drops in heron numbers seen in the graph seem to have a similar explanation, as they are also associated with cold winters. For example, the recent decline in numbers following on from 2010 was also associated with several cold winters (Fig. 126).

FIG 124. Changes in numbers of breeding Grey Herons *Ardea cinerea* in England and Wales, from the British Trust for Ornithology's ongoing survey (Massimino *et al.* 2019).

Such large cold-weather-related declines in heron numbers may become increasingly rare in future. Attempts to model the likelihood of such cold weather in Britain – driven by the interest of the insurance industry in the numbers of burst pipes – suggests that because of climate change the chance of a winter as cold as 1962/63 declined by a factor of two during the twentieth century and is likely to decline even more in future (Koumoutsaris 2019). The response

FIG 125. My mother climbing onto an ice floe on Southport beach (northwest England) early in 1963. (Lionel Wilkinson)

FIG 126. Dead Grey Heron *Ardea cinerea* in Lincolnshire during very cold winter weather at the end of 2010. This was one of the coldest winters in recent years, one measure of its severity being that the cost to the British insurance industry of burst pipes was in excess of £300 million (Koumoutsaris 2019).

of the heron numbers in the few years following these crashes is also instructive, as they tend to return to something similar to the population size before the decline. This is reminiscent of the logistic population model, with rapid growth at low numbers and a tendency to level off at higher population sizes. It suggests a carrying capacity for Grey Herons in Britain with some factor, or factors, regulating the population size.

These heron data illustrate two different types of processes which can affect the numbers of an organism, namely density-dependent and density-independent processes. The dramatic effect of a really cold winter on heron numbers is a classic example of a density-independent effect, which is something that can have a large effect on population size irrespective of the size of the population. While a cold winter reduces heron numbers, obviously there is no way that the numbers (or density) of herons can affect the likelihood of a cold winter – hence it is density-independent. This means that although winter mortality can have substantial effects on heron numbers it is not able to explain the apparent long-term stability in population size, as there is no feedback between heron numbers and the death (or birth) rate.

Extreme weather events, be they cold winters or summer droughts, provide good examples of density-independent processes, as the lack of feedback is easily appreciated in such cases (Fig. 127). They can be contrasted with

FIG 127. Collared Dove *Streptopelia decaocto* in winter. Cold winter weather, which can make foraging very difficult, is a classic example of a density-independent process in population ecology.

density-dependent processes, where there is feedback between the numbers of individuals and the factor(s) limiting population size, be that birth or death rate, breeding success or the tendency of individuals to move into (or out of) the population. In such a case there can be regulation potentially leading to a relatively constant population size from year to year. This is an important point. If it's unfamiliar, stop and think about it and convince yourself of the importance of this feedback in regulating the size of a population.

BIRD EXAMPLES OF DENSITY DEPENDENCE

A seabird colony can often be a good location to visualise the basic ideas behind density dependence. The great seabird cliffs of northern and western Britain are one of the most dramatic spectacles of British natural history during the breeding season, but they can often present a confusing sight. There are large numbers of birds, and it's usually difficult to get a good view of them – at least without access to a boat – to see what's going on. The Kittiwake colony in Dunbar harbour is smaller and more approachable, and so provides a clearer illustration of some of the principles of density dependence (Fig. 128). The birds breed on the

FIG 128. Kittiwakes *Rissa tridactyla* nesting at Dunbar harbour, East Lothian. The availability of good nest sites is one of the factors that limits the size and growth rate of a Kittiwake colony.

cliffs below the old castle walls, and on parts of the castle itself. At this smaller and more easily viewed colony it's much more obvious that there is a limit to the number of good nest sites. So as the colony grows some birds are going to be forced to try and nest on less good sites, and ultimately there may be no space for new nests at all. Indeed, studies of numerous Kittiwake colonies have suggested that competition for limited nesting spaces often limits colony growth. The extent to which this limits the total Kittiwake population will depend on the extent to which good sites for new colonies are available (Coulson 2011).

Just up the coast from Dunbar lies the Bass Rock, home to a major breeding colony of Gannets *Morus bassanus* – indeed, the specific name *bassanus* refers to the Rock, which historically provided a lucrative harvest of these birds for food. The Gannets have been here a long time, and are first mentioned in an account published in 1447 (Nelson 2002). During the last 50+ years the number of Gannets breeding on the Bass Rock has undergone a substantial expansion; for instance, in the 1960s most of the summit of the rock was free of nesting birds while today it's crowded (Fig. 129). This raises the question of what limits the size of a Gannet colony: is it just nesting space? One way to approach the question is to compare Gannet colonies of different sizes around the coasts of Britain and Ireland. Another approach is to look at changes at individual colonies over time as the colony size changes. The second approach requires historical data on the ecology and behaviour in the past, and for the Bass this is available because of detailed studies there in the 1960s by Bryan Nelson.

Sue Lewis and colleagues applied both of these two approaches to nine Gannet colonies around the British and Irish coasts and concluded that availability of fish was an issue. As a colony gets larger the time the parent birds spend on each fishing trip gets longer. On the Bass Rock between the 1960s and 2000 the length of a typical fishing trip doubled to around 18 hours. This clearly has implications for the parents' ability to feed growing chicks. The problem may be that the birds are reducing fish numbers to a point where there are few fish near the larger breeding colonies. Potentially the disturbance caused by a large number of birds feeding by plunge-diving into the sea causes the shoals of fish to scatter and/or swim deeper, making them less available. Indeed mathematical models of this process suggested that such an effect could explain the observed pattern in length of feeding trips (Lewis *et al.* 2001).

In the case of the Gannets, then, as the density of the colony increases it becomes ever harder to successfully feed the chicks, another nice illustration of density-dependent processes. This mechanism for limiting the size of a seabird colony was first suggested in the early 1960s – around the time Bryan Nelson was carrying out his PhD studies on the Bass Rock. The idea came from studies

FIG 129. Changes in numbers of Gannets breeding on Bass Rock from 1960s onwards. The top photo is from 1961, then 1986 (bottom), 2000 (following page top) and 2011 (following page bottom). This sequence of photographs effectively shows the logistic growth graph in pictorial form, with rapid population growth at low numbers eventually levelling off to a maximum value the Rock can support. (Photos by Bryan Nelson, except for 2011, which is by myself)

of tropical seabirds on the British Overseas Territory of Ascension Island by Philip Ashmole, and is now sometimes referred to as Ashmole's halo, for the hypothesised ring of fish-depleted water around the colony. By the mid-1980s a consensus was starting to emerge that many seabird populations were regulated by a mix of available colony sites, nesting sites within colonies and availability of food (Birkhead & Furness 1985). These conclusions still hold, and are now supported by better data than was available 35 years ago.

It's not just seabirds that provide instructive examples of density dependence. We have already seen how the harsh winter of 1962/63 affected herons mainly through a density-independent mechanism; other species were of course also affected, and we can turn to one of them for a further example of a density-dependent process. After suffering a threefold reduction in density as a result of the 1962/63 winter, the British Kestrel *Falco tinnunculus* population increased until around 1968, at which point it levelled off and stayed reasonably constant during the 1970s. An analysis of BTO data gave interesting insights into some of the processes involved in the post-1963 increase and why the rate of growth had slowed by the end of the 1960s (O'Connor 1982). At lower population sizes most birds were breeding in favoured habitats such as moorland and non-coastal cliff sites, but as the population increased a greater percentage began breeding in farmland or in coniferous woodland. As this wider range of less ideal habitats became used for breeding, the number of eggs laid in a nest tended to decrease; so as the population became larger more birds were forced to use less good habitats for breeding, and so the overall growth rate of the population declined.

Kestrels defend a small territory immediately around their nest site, which is one of the reasons why as the population increases many birds are forced to breed in suboptimal locations. In fact territorial birds often show good evidence for the importance of density-dependent processes. Experiments on a wide range of bird species in which territory holders have been removed show that in good habitat there is often a surplus of non-breeding birds ready to move into the gap and start breeding, and indeed sometimes birds will move in from poorer habitats to fill these newly vacant high-quality gaps (Newton 1992, 2013). Therefore territorial behaviour tends to promote density-dependent regulation in populations, by limiting the breeding density, and forcing some individuals into poorer habitat at higher densities.

Had this book been written around the middle of the twentieth century there would now follow a whole section discussing the question of whether populations are affected mainly by density-dependent or density-independent processes. Seventy years on, it's now clear that both can be important, and the relative importance of these two processes can differ between species and locations (but remember that only density dependence involves the feedbacks that can lead to population regulation). With the great benefit of hindsight it is clear that part of the reason consensus was slow to build was the different organisms people were studying. Scientists working on territorial birds saw extensive evidence for density dependence, while those working on invertebrates, especially ones living in climatically extreme environments, saw little other than density-independent processes. Some influential invertebrate ecologists even denied the existence of density-dependent factors at all (Andrewartha & Birch 1954). Certainly weather can have substantial impacts on invertebrate populations – think of the effects of a hot summer on Painted Lady butterflies in Britain, when such weather can allow several generations during the summer (Fig. 130). But density-dependent processes are often important as well. One area where this is the case is in mortality from disease, which is often greater in crowded populations.

FIG 130. The Painted Lady *Vanessa cardui* migrates into Britain each year from the south. Numbers arriving are in part determined by temperature and wind directions. Once in Britain, it takes around six weeks from egg to adult, so in a hot summer multiple generations can be fitted in, leading to years with large numbers of butterflies (Thomas & Lewington 2010).

PARASITES AND POPULATIONS

Parasite, in the way the term is often used in ecology, can cover everything from viruses to fleas, flukes and cuckoos (Rothschild & Clay 1952). The organism suffering from the parasite is technically referred to as the host. Since theoretical work by Roy Anderson and Robert May starting in the late 1970s it's become common to divide parasites into microparasites and macroparasites (Anderson & May 1991, Grenfell & Keeling 2007). Microparasites tend to include entities such as viruses, bacteria, protists and microfungi, while the macroparasites are the fleas, flukes and other usually tiny animals (the cuckoo strategy of laying eggs in other birds' nests doesn't really fit into this classification). Fundamentally, the main difference between micro- and macroparasites is reflected in the way that people usually model them using mathematics. Microparasites reproduce rapidly within their host, and normally they are described mathematically by keeping track of the number of infected hosts. Macroparasites on the other hand reproduce more slowly, and attempts to model them using maths usually try and follow the numbers of individual parasites.

In this section I am going to concentrate on disease-causing microparasites, as they can on occasion have major impacts on the host population size – especially when infecting a new population that has little, if any, immunity to them. High-profile examples of this include a number of recent problems with introduced tree diseases. For example, Ash trees are currently seriously affected by ash dieback caused by the fungus *Hymenoscyphus fraxineus*, which was first confirmed in Britain in 2012. This not only kills the trees themselves, but also indirectly affects lichens and other organisms which live on Ash trees and plants growing on the woodland floor below. As trees die, first more light gets through to the ground, affecting conditions for woodland-floor plants, and over the next few years this likely leads to an increase in the shrub layer, so slowly shading these forest-floor plants (Mitchell *et al.* 2016).

A more zoological example is the effect of squirrelpox virus – introduced with Grey Squirrels from North America – on native Red Squirrels, as described in Chapter 6. More recently another microparasite has been found in Red Squirrels from Brownsea Island in Poole Harbour; this is one of the leprosy-causing bacteria, *Mycobacterium leprae*. Genetic studies show that this is a very similar strain to the one that affected humans in Britain during medieval times, and it looks likely that it somehow passed from humans to squirrels or vice versa – currently the idea that humans infected squirrels seems the most favoured one. In addition, another leprosy bacterium, *M. lepromatosis*, has been found more widely in Red Squirrels in Britain and Ireland. It is not yet clear to what extent leprosy may be contributing to declines in the British Red Squirrel population (Avanzi *et al.* 2016).

When it comes to the effects of microparasites on populations, many of the most detailed examples come from humans, well studied because of the medical interest. For example, the influenza pandemic at the end of the First World War (caused by a virus) is thought to have killed in excess of 25 million people – many more than died in the war itself (Porter 1997), and I am putting the finishing touches to this chapter in late 2020, in the midst of the COVID-19 pandemic – which spectacularly shows that humans have not fully escaped from the influence of microparasites. In the fourteenth century the Black Death – caused by the plague bacterium *Yersinia pestis* – killed an estimated 30–50 per cent of the human population of Europe over a four-year period (Holmes 2011). Although probably the most famous, this was not the only outbreak of plague in British history, and there was another well-recorded outbreak in 1665. In London Daniel Defoe – the author of *Robinson Crusoe* – described the horror of it in his fictionalised *Journal of the Plague Year* (1722), writing 'Tears and lamentations were seen almost in every house … and death was always before their eyes'. It is an aspect of this later plague that I want to discuss in an ecological context.

Away from London, the place that figures most prominently in history books in relation to the seventeenth-century plague outbreak is the Derbyshire village of Eyam. Tucked up a little side road running up through limestone cliffs from the main Middleton Dale, the village is now an incongruous centre for plague tourism. During a midweek visit in February most of the visitor attractions were closed, with just a few tourists around the church. Inside is a finely illuminated copy of the register of deaths from 1665–66, while amongst the Yews *Taxus baccata* and other trees in the churchyard graves of plague victims mingle with those of centuries of other villagers (Fig. 131). The story of the plague at Eyam features not only in school history lessons and TV documentaries, but also in the technical scientific literature. Ecologists Roy Anderson and Robert May opened the introduction to their now classic 1991 book on mathematical models of infectious diseases in humans by describing the Eyam story.

Briefly, when the disease arrived in the village – supposedly via rat fleas in a consignment of cloth from London – the local rector with the support of the previous incumbent persuaded almost all of the villagers to stay put rather than risk spreading the contagion to other parts of Derbyshire. Over the next 15 months 257 people in the village died, out of a population of at least 700 people. If you consult older books on the Eyam plague these numbers will appear wrong; for example, at the time Anderson and May were writing, it was thought that the population was around 350, but subsequent historical

FIG 131. The church at Eyam. Inside the church is a register showing all the deaths from the Eyam plague of 1665–66. The tomb in the foreground is that of the rector's wife, who didn't survive the plague outbreak.

research has increased this estimate to the value of 700+ (Whittles & Didelot 2016). From a population ecology perspective, the interesting thing about Eyam is that almost no one entered or left the village while the plague ran its course – so we have a well-documented isolated population, which lends itself to mathematical analysis.

An early attempt to use mathematics to understand the Eyam plague was made by G. F. Raggett from Sheffield Polytechnic, who realised that the isolation of the villagers gave a closed population amenable to quantitative analysis (Raggett 1982). However, in what follows I rely mainly on the more recent work by Lilith Whittles and Xavier Didelot (2016), which used a more sophisticated approach, and up-to-date information from historians on the likely population of Eyam at the time. Plague is mainly found in rodents and often spread by rat fleas (e.g. *Xenopsylla cheopis*) which can sometimes also bite and infect humans (Fig. 132). However, person-to-person spread via human fleas and lice also happens, and it can also spread between people via aerosols following the development of pneumonia – so-called pneumonic plague. It's an interesting detail that the rats involved in the historic outbreaks of plague in Britain were not the common Brown Rat *Rattus norvegicus* (which didn't arrive in Britain until the eighteenth

FIG 132. Rat weathervane on top of the small museum in Eyam. It's likely, however, that human-to-human transmission of plague was more important than transmission by rat fleas in the village.

century) but the now very rare Ship, or Black, Rat *R. rattus* – which was introduced to Britain during Roman times.

Statistical analysis of the mortality data from Eyam suggests that around 65 per cent of the population survived the plague. If you rephrase that to say that over one-third of everyone in the population died over a period of less than two years then the full horror of the situation becomes starkly evident. Had you been living in Eyam at the time, your chance of survival was significantly affected by two things: (1) whether anyone else in your household was infected, and (2) whether you were poor, as surviving tax records show that – as is so often the case – those with less income suffered more. The point about prior infection in the household raises an important idea, developed in more detail in the next section, that spatial structure often really matters in population ecology. Early attempts to model the Eyam plague simplified the mathematics by assuming that within the village everyone mixed equally with everyone else, so the details of who lived where didn't matter to your chance of becoming infected. That was clearly not the case, a point brought home if you visit Eyam today by signs in the village displaying long lists of people who died in some households (Fig. 133).

As well as statistical analysis of the mortality rates, Whittles and Didelot (2016) used the historical data to construct mathematical models of the spread of plague in Eyam. These strongly suggested that there must have been a mix of rodent-to-human transmission and direct human-to-human transmission, with the latter much more important. There was also a seasonal effect, with fewer infections over winter; this could be because winter affected the behaviour of people and/or rodents. The maths suggested that the rat-related infections were more important over winter – so at least part of the answer was probably that people were meeting

FIG 133.
A visualisation of the importance of spatial structure in population ecology. The list of all the people from one Eyam cottage who died of plague, with their death dates. You were more likely to die if you shared a house with infected individuals. A similar pattern has been seen in the COVID-19 pandemic, with people living in more crowded accommodation being more likely to become infected.

and infecting each other less over the cold months. With over a third of this isolated human population dying over a 15-month period, the plague in Eyam is a striking example of the effect that disease can potentially have on population size. The role of spatial structure (in the Eyam case, that's living in the same house as infected individuals) has become a major topic in population ecology, including its application to nature conservation, and is considered next.

IT'S NOT JUST AVERAGES – PATTERNS IN SPACE MATTER IN POPULATION ECOLOGY

Up until now in this chapter the majority of examples I have discussed have considered that the population has a particular value of rate of growth (r), carrying capacity (K) or other characteristic that is the same across the whole population – although these may vary over time. It would seem very unlikely that some factor (such as germination success or breeding success) is identical at all points across a population's range, so this simplification is better seen as an average (e.g. average breeding success or growth rate etc.). This is a simplification that has often been assumed in population biology, but over recent decades there has been a big interest in variation in space and its importance both for understanding population ecology and for applying the results to nature conservation. The key term is 'spatial structure' – the idea that location matters and that factors such as population growth rate will vary in space as well as in time. This important idea has already been hinted at, in the impact of which house you occupied in Eyam in 1665 on your likelihood of dying of plague, and in the variation in habitat occupancy by Kestrels recovering from low numbers following the winter of 1962/63.

To get an idea of the importance of spatial structure, consider Wythenshaw Park in Manchester – a mix of open grass, trees and shrubs (Fig. 134). This looks

FIG 134. Wythenshaw Park in Manchester, photographed in 1989 at the time when David Groom (1993) was studying Blackbird *Turdus merula* breeding success, and the effects of Magpie *Pica pica* predation on the eggs and nestlings.

like reasonable Blackbird habitat, providing both feeding areas and cover for nest sites. In the late 1980s I was using the park as one of the study sites for my PhD work on urban trees – collecting data on the shade they cast to inform statistical models I was constructing to help describe and predict the shade trees cast around buildings. Unknown to me at the time, there was another PhD student, from a different university, using the park as one of his field sites. David Groom was doing a PhD at Manchester University, supervised by Derek Yalden, looking at the effects of predation on Blackbird nests. Groom studied the breeding success of Blackbirds at Wythenshaw and two other parklands in southern Manchester. What he found was a spectacularly low breeding success: at Wythenshaw only around 4 per cent of nests fledged young each year (a more typical value for sites in Britain was anything from 28 to 44 per cent). However, my memories of the park at the time contain plenty of Blackbirds, and Groom's more quantitative data also showed no shortage of the birds. This raises two interesting questions – where were the Blackbirds coming from if breeding success in the park was so poor, and why was the success rate only around 4 per cent?

The main reason for the very low breeding success appeared to be predation by Magpies. It's often not easy to establish which predators are raiding a nest – especially in the 1980s before remotely operated 'trail' cameras became common – so in this study plasticine eggs were added to nests and beak marks in these eggs showed that Magpies were the main predator. Groom concluded that the very high mortality rates must have meant that the Blackbird population in the park could not have been self-sustaining and was only surviving because of birds moving in from surrounding areas where breeding success was higher. The term used to describe a population like the Blackbirds in the park, which can only survive with the aid of immigrants because of low local breeding success, is a 'sink population', while the areas with better reproduction rates which are supplying the immigrants are 'source populations' (Pulliam 1988, Nee 2007).

In British history the big industrial cities of the nineteenth century could be so unhealthy that they were sink populations too, with mortality rates such that the city relied on immigration from the wider country to maintain its population. Take Liverpool as an example. Because so many of its population in the mid-nineteenth century were employed around the docks they were often particularly poor, as many were casually employed, paid only when needed. As with the plague in seventeenth-century Eyam, it was as usual the poor who suffered most, living crammed together in insanitary conditions. In the 1840s the average age of death in Liverpool was just 19 (Chave 1984). For almost a quarter of a century I lectured at Liverpool John Moores University, and the big first-year ecology classes could have over 200 students in them. In the mid-nineteenth century over half the people in the lecture theatre would have been dead by the time they were of undergraduate age if they were a random sample of the people of the city, and imagining a large lecture theatre half-full of ghosts was a striking image of the extent of this mortality! Infectious diseases such as typhoid and typhus (both bacterial), and a long list of others, played a large part in the high death rate. The impact of these diseases was so bad, and so obvious, that in 1847 Liverpool employed William Henry Duncan as the first medic in a public health role in Britain to try and alleviate the problems. It's primarily a combination of improvements in sanitation and other living conditions, along with vaccination, which has stopped cites such as Liverpool remaining population sinks.

The concept of source and sink populations is one key spatial idea in population ecology, and it is closely related to another important idea, that of metapopulations (Hanski 1998, Nee 2007). Consider a collection of ponds scattered across an area of agricultural land. Small ponds can easily dry up from time to time, or be made unsuitable for some species by events such as pollution by agricultural chemicals. So we can imagine a set of ponds that provide home to a particular species – for example, various dragonfly nymphs – but which ponds are actually occupied at any one time changes from year to year. Providing the species can (re)colonise new ponds, it can persist in this landscape, despite repeated extinctions in any one pond. This is an illustration of the basic idea of a classic metapopulation – often called a Levins metapopulation after Richard Levins, who introduced the term in 1970. So more formally a metapopulation is a population of populations – a collection of local populations linked together by dispersal.

The hypothetical network of ponds I have just described is actually based on a real study of dragonflies and damselflies in a group of 11 ponds in rural Cheshire studied during 1997 – although in this case they were not prone to drying up but did suffer pollution events from agricultural chemicals. This study tracked the movement of marked adults of seven species of dragonflies and damselflies between the ponds (Fig. 135), showing that although most individuals stayed by their original ponds, a few moved to more distant ones – so there was the potential for ponds to be (re)colonised by any of these species if they became extinct in a particular pond. If colonisation and recolonisation of ponds is important to the long-term persistence of the wider metapopulation, then the extent of dispersal between ponds becomes a crucial factor for its conservation – this was the rationale behind the Cheshire pond study of the movement of marked insects. Indeed, the opening sentence of the summary of the research paper describing this study read 'Dispersal is an ecological phenomenon which is of fundamental importance to population biology' (Conrad *et al.* 1999). So a metapopulation

FIG 135. The Common Blue Damselfly *Enallagma cyathigerum* was one of seven species studied in a network of Cheshire ponds by Conrad *et al.* (1999). It's not just the insects that are moving between ponds, their parasites are too. The two small red spots on the underside of the abdomen of this damselfly are parasitic water mites.

approach focuses attention on dispersal, along with extinction at the level of individual habitat patches, such as ponds in the case just discussed (Fig. 136).

Insects often feature in metapopulation studies; it seems that this is an idea well suited to many insect populations of conservation concern, with butterfly studies being particularly prominent. The classic example has become work by Ilkka Hanski and his collaborators on the Glanville Fritillary *Melitaea cinxia*. In Britain this is a rare butterfly restricted to a few places in the south – notably the Isle of Wight and the Channel Islands – but it's much commoner in Finland where Hanski studied it in the Åland Islands. Here the larger metapopulation survives across a range of patches of suitable habitat, but each year the butterfly becomes extinct in some patches while others are colonised. A nice autobiographical account of this work was given by Hanski in his last book, which he wrote while dying of cancer. In it he described how as a student he became interested in the metapopulation ideas of Levins and decided that butterflies would be a good group to use to study these, eventually deciding on the Glanville Fritillary (Hanski 2016). Edward Wilson (1992) memorably likened this type of metapopulation to 'a sea of lights winking on and off across a dark terrain', with each bulb a potential habitat patch, which is lit when occupied and dark when empty.

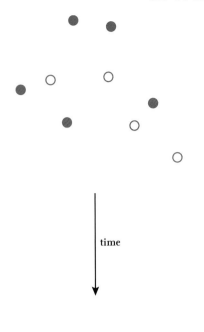

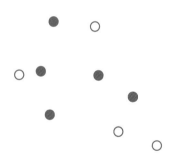

FIG 136. The basic idea of a Levins metapopulation model. The circles are habitat patches (such as ponds) which can be either occupied (●) or empty (○). The diagram shows the imaginary results of two surveys separated by time. In both cases five ponds are occupied, but some ponds have been colonised while others have lost the species, and some have maintained the species between the two sampling dates. The arrangement of ponds in the diagram is based on Conrad et al. (1999), except in their case the top-right point was made up of three smaller ponds all very close together. Note that this study only looked at movement between ponds in one year, rather than changes in occupancy over several years, as in this hypothetical example.

However, metapopulation ideas can apply to a much wider range of examples than the commonly cited butterfly studies. Think of humans and common cold viruses – we are habitat patches for the virus, and over the years can be colonised and recover (i.e. the virus becomes extinct in 'our' patch) on multiple occasions. Indeed, much of epidemiology is really metapopulation ecology. Alternatively, visualise an orchard with windfall apples littering the ground. These apples are habitat patches for a range of yeasts and other microbes, with the added complication that the patches regularly become extinct as the apples rot away. Something similar can happen in the epidemiological cases if a host becomes immune when recovered, and so is no longer available as a 'patch' in the system.

The implications of thinking about populations in these spatial ways are rather interesting, and initially can seem counterintuitive. Sean Nee (2007) made the point eloquently, writing that:

> *Species are to be found wherever there is suitable habitat that they can get to. They will be most abundant in their preferred habitat and species can be preserved as long as a good-sized chunk of suitable habitat is conserved for them.*

He then asked whether, in the light of ideas of sources and sinks and metapopulations, the statements 'still sound reasonable?' In the examples I have just described, suitable habitat (e.g. a pond or a patch of butterfly breeding habitat) may at any point in time be empty of the species of interest, but that habitat patch may still – through (re)colonisation – be an important part of what is required for the metapopulation's long-term persistence.

Or consider Wythenshaw Park in the late 1980s – it looked good Blackbird habitat, and indeed there were plenty of Blackbirds. However, as a sink population it was playing no role in the continued survival of Blackbird populations in southern Manchester. If the breeding success of a population declines for any reason, then any source populations may be producing far fewer individuals, leading to fewer immigrants and therefore range contraction as the sink populations vanish when they are no longer receiving enough subsidies from the sources. This could happen without any obvious habitat changes, and no amount of preservation of the habitat at a site will help if continued persistence relies on influx of individuals from the declining source populations elsewhere (Lawton 1996).

CONCLUDING REMARKS

The Swifts in Selborne illustrate both the types of questions raised in population ecology and the importance of long-term data sets in trying to answer them. As in the Sherlock Holmes story where the key clue was that a dog didn't bark in the night, Gilbert White realised that the absence of a big increase in Swift numbers was telling him something profound. Biological populations have the potential to grow rapidly by doubling every generation – exponential growth – and yet in most cases the numbers of a species that we see from year to year stay relatively constant. Why is this? In cases such as the Selborne Swifts, or English and Welsh Grey Herons, where we see a long-term constancy in population size, unless we put it down to just chance (effectively the same as always getting 'tails' every time we toss a coin) then there must be something that tends to control numbers. Something operating a bit like a thermostat in a heating system.

If you have a thermostatically controlled heating system then you need to set it to a particular temperature. In population ecology this setting is the carrying capacity, the maximum number of any species that the environment can support long-term. Logically it's obviously correct that there must be some such limit, although it's often difficult to define this exactly – for example, it can vary from year to year (or from week to week for short-lived organisms). For a population to level off around the carrying capacity there needs to be some sort of feedback between population size and factors that affect population growth (such as birth or death rates, or likelihood of immigration), so that there are higher death rates or lower birth rates at larger population sizes. Factors such as very cold winters can greatly affect population size, but cannot have this regulating effect because of the lack of such a feedback mechanism.

One example of a potential density-dependent mechanism is the effect of crowding on the likely spread of microparasites (diseases). On occasions these can lead to very substantial declines in population size, and there are good examples of this from human history such as the Black Death or the flu pandemic at the end of the First World War. Between the writing of the first and final drafts of this chapter the fact that disease can still profoundly affect human populations has been made plain by the COVID-19 pandemic. And in his poetic response to these recent events Simon Armitage, the British poet laureate, drew on another of the examples used in this chapter – the Eyam plague of 1665–66 – in his poem 'Lockdown'.

Population ecology matters for many reasons, including the fact that it's the background theory to understanding things that can kill you. Questions about vaccination rate, such as what percentage of the population must be vaccinated to prevent a disease spreading and becoming epidemic (herd or population immunity), are population ecology questions.

The big growth area in population ecology in recent decades has been in understanding the importance of spatial structure. This is important for basic theoretical understanding, but also for applications to nature conservation or epidemiology. The key ideas have been source/sink dynamics and metapopulation theory. These can have counterintuitive results, such as the idea that habitat patches where you currently *don't* find an organism can be key to the long-term survival of that species. So although populations can vary considerably over both time (think of the heron data) and space (think of the Blackbirds in Wythenshaw Park compared to other more successful source populations), there are a number of processes which tend to limit the extent of this variation. In many cases long-term stability may only be a realistic prospect at the metapopulation level, with extinctions repeatedly happening in individual local patches.

CHAPTER 9

Succeeding in Wicken Fen

The view in front of me is a bit depressing. On an afternoon in late
August I am looking out over the northern part of Burwell Fen in
Cambridgeshire, and in front of me lies rather nondescript former
agricultural land with a few grazing Highland cattle. Prior to large-scale drainage
in the seventeenth century this area of the Fens was rich in wetland wildlife,
and in fact this part of Burwell Fen wasn't fully drained until during the Second
World War. There is a hint of optimism in the view too – the fen was returned to
conservation management in 2001, although there is a long way to go to return to
the biodiversity that would have been here before drainage. And that is the theme
of this chapter: how the vegetation at a site, and the assorted other organisms,
change over time. In more technical language, this chapter is about ideas of
'ecological succession'.

Behind me, to the northwest, things are rather more encouraging for a
naturalist, as here more of the wetland habitats survive on parts of the National
Trust reserve at Wicken Fen, which has been under conservation management
for longer. I have spent the morning on the sedge fen on this reserve and,
having taken the photographs I wanted of the remains of Burwell Fen (Fig. 137),
I head back across a footbridge over the Burwell Lode, part of the network
of waterways that helped drain the Fens. The afternoon is hot and my shirt
damp with sweat under my rucksack full of cameras. Along from the bridge,
three teenage lads have just arrived on the riverbank and have stripped for
swimming – an almost bucolic scene that could have been from the early
twentieth century, before the last of Burwell Fen was drained, if it were not for
the loud music they have brought with them. I head north along a cycle track,
on my left wetter habitats managed for conservation and to my right duller

FIG 137. Burwell Fen photographed in 2018 from near Cockup Bridge. Until the 'Agricultural Authority caught, cowed and drained it' during the Second World War most of this view (except a few fields on the left-hand side of the photograph) were still fen, rich in natural history (Ennion 1949). It is now being managed for nature conservation.

agricultural fields, enlivened by a skein of geese flying over them. Fence posts along the track are splashed orange with *Xanthoria* lichens, attractive but likely a sign of high levels of nutrients drifting off the agricultural fields. On reaching the clear waters of Monks Lode, all dragonflies and small fish, I turn left and head back to the Wicken Fen visitor centre.

Wicken Fen is one of the oldest nature reserves in the country, dating to the very end of the nineteenth century (Fig. 138). The first parts of the reserve were acquired by the National Trust on 1 May 1899, and over the next few years several entomologists sold or donated land to the Trust which they had bought to preserve its natural history interest and protect their insect collecting sites. This gives Wicken an important place in the history of nature conservation in Britain, 'one of the oldest and most revered of all Britain's great nature reserves' (Pryor 2019). Being so near the University of Cambridge, the reserve is also an obvious site for ecological research and teaching, and it's within cycling distance of Cambridge if you are reasonably fit – an important point in the early twentieth

FIG 138. Wicken Fen – the view east across Drainer's Dyke in 2018. Most of the land visible in this photograph was donated to the National Trust at the start of the twentieth century. The book edited by Laurie Friday (1997) provides a detailed history of the reserve up to the end of the century.

century before the car became so ubiquitous (in those days you could cross the river at Upware by boat).

In 1927 Harry Godwin set up what was to become a famous long-running experiment on the Fen; parts of this experiment are still running almost 100 years later. It is these experiments that provide a good introduction to the idea of ecological succession. To find them from the National Trust visitor centre, walk past the much-photographed small windmill (a scoop wind pump, of the sort that used to be used in fen drainage before the rise of steam pumping technology in the nineteenth century) and carry on anticlockwise along the path running round the reserve. Just after a left turn when you reach Drainer's Dyke – which dates back to early seventeenth-century drainage – you will find the experimental plots marked out on your left by a rope barrier and a display board explaining their importance.

OF PLOTS AND TRIANGLES –
GODWIN'S CLASSIC EXPERIMENTS

The first plot you reach is now woodland with birch trees, followed by four linked plots of more grassy and herbaceous-looking vegetation. At first glance it looks as if there are just two experimental treatments – the wooded end plot plus a large expanse of Common Reed and similar plants. But look more closely, and more subtle botanical differences can be seen as you walk the length of the experiment. Walking along the path away from the wood, you see a more diverse vegetation, with sedges and some Purple Moor-grass *Molinia caerulea*, while towards the far end the vegetation becomes simpler, with increasing amounts of reed. These differences in vegetation are the result of over 90 years of experimental management, albeit with a break in the experiment from 1940 to 1954 (Fig. 139).

The background to Godwin's experiments is that the historic management of much of the Fen had been through cutting, mainly to provide a crop of Great Fen-sedge *Cladium mariscus* for use in thatching, as well as for fuel and a variety of other

FIG 139. The Godwin Plots at Wicken Fen, photographed in August 2018. The trees that form a wall beside the path are growing on the plot that has been left uncut since 1927, while the other plots have been cut on cycles of every one to two years (nearest the camera) to every four years (next to the wooded plot).

purposes. This management had been used at the site for at least 600 years, and had created much of its conservation interest. However, extensive cutting stopped once Wicken Fen became a nature reserve, a change which accidentally may have been 'the most significant and damaging' one in the history of the sedge fen (Rowell 1997).

Godwin's experiments were designed to look at the effects of cutting on the fen vegetation, in part as an academic exercise to understand succession in fenland systems, but they also produced results that helped inform conservation management. In 1927 Godwin set up two sets of experimental plots, one in sedge fen and the other in 'litter' – a fen type dominated by Purple Moor-grass (Friday *et al.* 1999).[4] The 'Godwin Plots' you can see today at Wicken are the sedge-fen ones; the litter plots were further along the path but were not reinstated in the 1950s. In both cases five 20 × 20 m plots were set out in a row with different frequencies of cutting. One plot in each set wasn't cut at all, while the others were cut in autumn – either every year, or every second, third or fourth year – with the cuttings removed. The autumn cut date was because Godwin thought that this was the traditional time of year to cut sedge on the fen, but later historical work by Terry Rowell (1997) showed that in fact this cutting was usually done in early summer. The ecology of Great Fen-sedge was studied in detail at Wicken by Verona Conway, a research student of Godwin's in the 1930s. She showed that most of its growth was over the summer, so that Godwin's late cut meant that there was mainly frost-sensitive young growth left exposed over the winter, which meant that the sedge struggled to survive frequent cutting at this time of year (Conway 1936).

Godwin ran the experiments from 1927 until 1940, and the plots were then abandoned and slowly turned to scrub. However, in 1955 the sedge plots were restored to use as a teaching aid for Cambridge students under the direction of David Coombe and Peter Grubb. The plot that had become scrubby woodland (dominated at the time by Alder Buckthorn *Frangula alnus*) was left uncut and the regular autumn cutting cycles were reinstated on the other plots. This means that the woody plot at the left-hand end of the experiment hasn't been cut since 1927, so if you leave sedge fen uncut for over 90 years woodland develops.

Typically, in designing experiments, you build in replication, that is multiple examples of the same experimental treatment to see if they all respond in the same way – although this is obviously easier for lab-based experiments than it is for large-scale field experiments. The second set of Godwin Plots no longer exists, as they were not reinstated in the 1950s – however, follow the path past the sedge plots until it takes a turn to the left at the junction of Drainer's Dyke

---

4 Friday *et al.* (1999) give a much more detailed account of these experiments than my short summary, along with references to Godwin's original publications on the experiment.

and Wicken Lode and you will find a second example of an experiment where open fen has turned to rather scrubby woodland, just to the right of the path. In 1923–24 a small triangle of fen was set aside, again by Harry Godwin, to follow fenland succession when cutting has been abandoned, an area now known as the Godwin Triangle (Fig. 140). This part of the fen had been cut for sedge during the First World War, and a rather poor-resolution photograph taken in winter 1928 (Plate 1 in Godwin *et al.* 1974) shows open fen with a scatter of a few small shrubs – unidentifiable in the photo but the figure caption suggests they were mainly Alder Buckthorn, Grey Willow *Salix cinerea* and birches.

Godwin surveyed the triangle, including plotting the position and size of all shrubs, in 1923–24, 1929 and 1934, with a more limited survey in 1944, by which time the triangle was better described as shrubby woodland rather than fen. After this Godwin's interests moved on to other topics, and it wasn't until 1972 that he did his last survey of the triangle. By this point he was in his early seventies and so co-opted two Cambridge undergraduate botany students to do the fieldwork – Brian Huntley and Debbie Clowes. Brian went on to an academic career (and

FIG 140. The Godwin Triangle (left of the path) at Wicken Fen, photographed in August 2019. This was an area of open fen with a scatter of small shrubs in 1923–24 when it was set aside as an experiment in succession by Harry Godwin. It has been left uncut since this date.

supervised my Masters research in the mid-1980s) and Debbie Clowes became a school biology teacher. Brian remembers that Sir Harry Godwin turned up in one of their university practical sessions and asked them if they wanted to help over the summer, although he can't remember why they were approached (he guesses Peter Grubb or David Coombe recommended them). That August Godwin took them out to Wicken Fen and showed them 'how to identify with confidence the principal species, especially the woody species, and how to map the individuals in a way that was consistent with his earlier surveys'. They stayed at the Fen for several days collecting the new data, which Godwin then analysed in the context of his previous results. The following year they were shown the draft manuscript, and Brian remembers:

> We had the chance to read it before he submitted it, but as you might guess as undergraduates working with a knighted professor, we were pretty unlikely to do more than say 'great' and be flattered to be included as authors.

The results showed that following the cessation of cutting there was an initial dense colonisation by Alder Buckthorn, in which Grey Willow played a rather transient role, and that later 'normal' Buckthorn *Rhamus cathartica* came to play a more important role, mainly via the expansion of existing bushes (Godwin *et al.* 1974).

The Godwin Plots and Triangle show that if you stop cutting the sedge fen then scrub and woodland develop. However, comparing the uncut plot with the triangle (Figs. 139 & 140) shows that although they have been uncut for the same period of time the look of the triangle is still rather 'scrubby' while the plot is better described as woodland. So clearly, even on the same reserve, site-specific differences can lead to somewhat different outcomes. However, in both cases the uncut vegetation has succeeded towards woodland. This has obvious applied importance for conservation. To maintain the sedge fen, regular cutting is required – on the reserve outside the plots this is done over the summer rather than in the autumn (Fig. 141).

In the context of climate change, people regularly discuss the widespread use of tree planting to remove carbon dioxide from the atmosphere, but often in Britain just leaving land unmanaged will lead to woodland without having to actually plant the trees. I will return to the more applied aspects of ecological succession later in this chapter, but first let's examine the main types of succession and then the ecological processes that cause these changes. I will return to Godwin's experiments as a source of data to illustrate key ideas, as well as discussing a range of other examples.

FIG 141. Sedge Fen at Wicken just after cutting in August. Summer cutting of the fen is an important part of its management, as it helps maintain the sedge fen and prevents it succeeding to scrub.

DESCRIBING SUCCESSION, AND THE NATURE OF THE CLIMAX

Clearly, if you remove the regular disturbance caused by cutting, or potentially grazing in a less managed system, then sedge fen tends to turn into woodland on a timescale of a few decades. This type of change in a plant community over time is usually called 'succession'. A more general definition of ecological succession is that it's the way biological communities reassemble following natural or human disturbance (Chang & Turner 2019). These ideas were first developed at the end of the nineteenth century, especially by Henry Cowles, who was studying vegetation on sand dunes around Lake Michigan in the USA – where, as the lake water receded, newly exposed ground was being colonised by vegetation. Similar ideas about coastal sand-dune succession were developed by several people in the late nineteenth century – including Eugen Warming (1909) in Copenhagen. The idea developed that the vegetation passes through a series of stages to eventually reach an endpoint determined by climate, which was assumed to be largely stable. For it to count as succession in this sense the changes should be caused by the effects of the vegetation, for example on the soil, rather than by external factors such as changes in climate (Wilson *et al.* 2019).

The basic idea is nicely illustrated by Harry Godwin's interpretation of succession in fen vegetation, based on his studies at Wicken Fen (Fig. 142). He suggested that shallow open-water vegetation is colonised by reeds, and that over time the dead reeds make the water progressively shallower until sedge takes over. This is then invaded by shrubs, first Alder Buckthorn and then normal Buckthorn as seen in the Godwin Triangle experiment. Over time this scrubby woodland develops into more mature woodland, possibly dominated by oak – but this was a guess, as this proposed final stage was missing from Wicken Fen. The basic story in this case is of the site becoming less wet over time, in part at least because of the effects of the vegetation, and woody plants becoming more dominant, with deciduous woodland being the final endpoint dictated by the Cambridgeshire climate (the so-called climatic climax). Godwin considered that regular cutting of the reeds or sedges deflected the succession away from its trajectory towards woodland, creating the mixed sedge fen or litter fen (with more regular cutting) that were historically important habitats at Wicken. Litter-fen species favoured by very regular cutting included Devil's-bit Scabious *Succisa pratensis* and Early Marsh-orchid *Dactylorhiza incarnata*.

Godwin's feeling that deciduous woodland is the likely endpoint of successional processes in many parts of Britain is often correct. Away from the wet conditions of the Fens another long-term experiment illustrates this nicely. Rothamsted Experimental Station in Hertfordshire is the home of many of the longest-running ecological experiments in the world (Chapter 11 provides a more

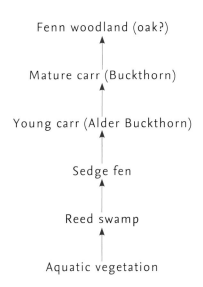

FIG 142. Scheme showing postulated succession at Wicken Fen, based on Godwin (1978). Cutting of reed swamp and sedge fen prevents the succession to carr (wet woodland) vegetation. As described in the text, there is evidence to suggest that at a wetland site such as Wicken succession may tend towards *Sphagnum* bog rather than the drier oak woodland postulated by Godwin and others. The various stages in a succession are often referred to as seres – so the succession pictured here would have six seral stages.

detailed description of Rothamsted). Two of these are very relevant to succession and woodlands – namely the Broadbalk Wilderness (Fig. 55) and the Geescroft Wilderness. Both of these areas are now covered by small woodlands but were agricultural fields until around the start of the 1880s (in the case of Broadbalk it's a rather thin woodland that could better be thought of as a very thick wooded hedge). They were then left uncultivated and the changes in the vegetation monitored periodically, and both eventually became variants of Ash woodland (W8 in the British National Vegetation Classification: Harmer *et al.* 2001).

However, it's likely that at Wicken Fen things might be a little different, with woodland not the final endpoint of succession. Godwin's assumption that woodland should be the endpoint was (and in many cases still is) widely shared, but on peatland like Wicken this seems often not to be the case. In an important study published back in 1970 Donald Walker used the fact that peat can record the vegetation history of a site – through preserving pollen, leaves, seeds, etc., in the sediment – to look at what really happens over timescales too long for even the most long-lived ecologist to study directly. In fact the book he published this work in was a collection of essays 'in honour of Harry Godwin'. He looked at 20 such studies from Britain and found that the actual course of succession was more variable than shown in Figure 142. In many cases wet woodland did form, but it was often invaded by bog moss *Sphagnum* spp., and eventually the trees died out and open raised bog with *Sphagnum* appeared to be the endpoint of the succession (see Chapter 10 for more on the effects of bog moss). Similar results were found for North American sites too (e.g. Klinger 1996). Indeed it's possible that raised bogs may have been present in the Fens in the past but all evidence for them has been lost due to drainage and cutting of peat for fuel (Godwin 1978).

In the original version of ideas of ecological succession these endpoints, be they deciduous woodland or raised bogs, were thought of as 'climatic climaxes'. In the view of one of the pioneers of succession theory, Frederic Clements, discussed in an influential article published in the *Journal of Ecology* (1936), 'The relation between climate and climax is considered to be the paramount one'. The idea was that any location will have a climate that is appropriate for a particular vegetation type, and that successional processes will continue until the vegetation type is in equilibrium with the climate (the so-called 'climax' vegetation – and yes, people have commented on the somewhat Freudian nature of the term, and over the years generations of students have introduced assorted double entendres into their answers to exam questions on succession).

There is a problem with this idea of succession to climax that is more obvious today than it was to the original proposers of these ideas. We are far more aware of the changing nature of climate than were most people 100 years ago – that is, both changes in the recent past and potential human-driven changes in the future. For example, in the 1890s when Cowles was studying succession in Michigan the Thames in London partly froze in the winter of 1895 (Kington 2010). The nineteenth century was the tail end of the 'Little Ice Age', so a sapling recorded in the first census of the Broadbalk Wilderness experiment in 1886 will have germinated under rather different climatic conditions than those it now experiences (Fig. 143). Because of this it seems more sensible to write about 'succession towards climax' than 'succession to climax' – as climatic change makes any hypothetical climatic climax at best a shifting target (Wilson *et al.* 2019). Although less of an issue in Britain, as our soils tend to be geologically recent because of the effects of the ice sheets of the last glaciation, succession can extend beyond climax as eventually depletion of soil nutrients causes further vegetation change. The volcanic islands of Hawai'i provide a classic example of

FIG 143. Man dragging firewood across a frozen lake, a tail-piece illustration from Thomas Bewick's *History of British Birds*, published at the end of the eighteenth century during the final phase of the 'Little Ice Age' when winters were colder, and glaciers were advancing in the Alps. Indeed, during the 1830s, conditions were close to those required to start glacier formation in the Scottish Highlands (Kington 2010). The classic white Christmas of Charles Dickens' novels is likely influenced by the fact that, of the first nine Christmases of his life (1812–1820) in London, six were white with either frost or snow (Lamb 1982). The key point for this book is that trees starting to grow at this time experienced as saplings a climate rather different from what they experience today. This undermines the old idea of a stable climatic climax.

this, with soils ranging from very recent (forming on recently erupted deposits) to over 5 million years old on islands that are no longer volcanically active (Vitousek 2004). Such a view of succession has less certainty, and more scope for variable outcomes, than the more deterministic early versions of this idea.

These ideas of ecological succession are normally applied to changes happening over tens to a few hundred years, but rather similar changes also take place over longer timescales of 1,000+ years (Birks 2019). A nice example is Sydlings Copse, a small woodland a few kilometres east of Oxford. Tufa deposits (tufa is a limey carbonate-rich soft rock formed under some freshwater conditions) have preserved both pollen and snail shells and allowed the long-term woodland succession at the site to be reconstructed (Preece & Day 1994). Around 9,300 years ago Hazel *Corylus avellana* woodland started to replace more open conditions. This unshaded marsh, prior to woodland development, contained snails such as *Vertigo angustior* and *Discus ruderatus* – this latter species is now extinct in Britain and is mainly found in more eastern parts of Europe. As the Hazel woodland developed, around 8,990 years ago, *D. ruderatus* was replaced by *D. rotundatus* (now a very common snail of British woodlands). Later – around 7,180 years ago – Small-leaved Lime and Alder started to become common in the woodland; the snail *Oxychilus cellarius* first appeared at this time. Over the 4,000 years recorded in this tufa deposit, as the vegetation changes so do the snails; however, there are also some changes in the snail community at times when the vegetation appears stable. Clearly over such a time period it is even less realistic to exclude the effects of a changing climate (and other factors) than it is with more conventional ecological successions happening over a timescale of tens to hundreds of years. So the idea of a static climax is even more unrealistic. We are clearly looking at a dynamic system rather than one settling into equilibrium with fixed conditions, and timescale can be important. What's stable on a year-to-year timescale may look less stable over 100 years, but may possibly look more stable when viewed from a distance over a large span of geological time.

TYPES OF SUCCESSION – PRIMARY AND SECONDARY

Succession is normally classified into two main types, primary and secondary. Primary succession starts on bare ground with little, or no, life already there. Secondary succession starts on ground that already has vegetation, or at least seeds buried in the soil, usually following some sort of disturbance to the previous vegetation. This disturbance may be natural, such as fire, storm damage, or following a landslip or avalanche (Fig. 144); or it may be due to

FIG 144. Broken trees melting out of avalanche debris near the summit of Rochers de Naye, above Montreux in Switzerland. An example of a natural disturbance event that will lead to secondary succession as the woodland returns to the slope following the removal of trees by the avalanche.

human management – such as mowing or burning. The main examples used so far in this chapter – of succession following the cessation of sedge cutting at Wicken Fen and the colonisation of former agricultural land by woodland on Rothamsted's Broadbalk and Geescroft wildernesses – are all examples of secondary succession. In the context of the definition of succession used above (the way biological communities reassemble following natural or human disturbance), in the case of primary succession the disturbance is so severe that no life survives it, and the ground has to be totally recolonised – as would have been the case over much of Britain following the retreat of ice sheets at the end of the last glaciation.

The classic textbook examples of primary succession tend to be the colonisation of recently erupted volcanic deposits, ground appearing from under retreating glaciers, or sand dunes – as in the early studies by Henry Cowles. With the exception of sand dunes these are examples you can't currently see in Britain; for retreating ice you would need a time machine that could transport you back 13,000 years to the end of the last glaciation, or more practically a trip to the Alps, while for erupting volcanoes your time machine would need to have a considerably longer range to travel back to the west of Scotland around 55 million years ago.

The obvious way to study how vegetation changes as a glacier retreats – or on newly formed volcanic land – is to sit and watch. The problem is that such an approach is extremely time-consuming, and ecologists just don't live long enough to study many primary successions this way. The often-used alternative approach is to substitute space for time, so that as you walk down an Alpine valley away from a retreating glacier you are on land that emerged from under the glacier longer and longer ago. So several hundred metres down-valley you may be on land that was under the ice a decade or two ago, while after a kilometre or so you may be on land that has been ice-free for a century. If you describe the vegetation along your walk then you are looking at 100+ years of succession without having to invest in a time machine. This approach is usually called a chronosequence study, and makes an obvious assumption that nothing else is affecting things other than the time since the ground was first colonised by plants. In the real world, of course, other things are changing too – for example, climate has changed since the nineteenth century, as described above. This means that chronosequences have to be interpreted with some care, and this is one reason why several of the key examples used in this chapter (e.g. the Godwin Plots/ Triangle and the Rothamsted wildernesses) are not chronosequences but sites where we have real, rather than inferred, data on how the vegetation has changed over 100 years or so. Where it has been possible to compare a chronosequence reconstruction with actual data collected over time, the chronosequence approach has often been found to be somewhat misleading (Johnson & Miyanishi 2008).

Despite the uncertainties with a chronosequence approach, it's still often used where alternatives are lacking. In the absence of retreating glaciers, the British habitat which appears to lend itself best to this approach is probably sand dunes. Imagine standing on a beach looking at whichever sand-dune system you know best – in my case it's the sand dunes along the Sefton coast north of Liverpool, which at around 21,000 ha is the largest dune system in England (Smith 2009). At the back of the beach, blown sand is starting to accumulate into small piles making embryo dunes, accreting around the very limited number of plants that can survive in such hostile conditions, particularly the grasses Sand Couch *Elyma junceiformis* and Lyme-grass *Leymus arenarius*. Behind these, larger sand dunes build; these are mobile dunes with often plenty of bare sand which can move with the wind, but are partly stabilised by another grass, marram (Fig. 145). Behind the mobile dunes things start to become more stable on so-called fixed dunes, with a more diverse collection of plants which can include more woody species such as Sea-buckthorn *Hippophae rhamnoides* and Silver Birch *Betula pendula*. It's not just the plants but the soils that are changing along this succession. Mobile dunes are sometimes called 'yellow dunes' from the colour of the exposed sand (Fig. 145), while the fixed dunes are referred to as 'grey dunes' as organic matter from dead plants has started to darken the colour of the sand. Interpreting all this as a chronosequence suggests that at some point in the past the fixed dunes were mobile ones, and even earlier they were embryo dunes, and before that blowing beach sand. As a broad generalisation about sand-dune systems this is probably

FIG 145. Mobile (yellow) dune at Ainsdale National Nature Reserve on the Sefton coast, north of Liverpool, with a mix of Marram *Ammophila arenaria* and bare sand. As an illustration of the difficulties in using the chronosequence approach to studying succession, many other factors have changed over time at this site. For example, in the 1960s, when I was a child, disturbance by human trampling was much more extensive – many kids, including myself, ran up and down the sand dunes, so that bare sand was much more extensive than is now the case (see also Smith 2009).

FIG 146. Lichens colonising a bare rock surface near Capel Curig, Snowdonia. Species include *Parmelia sulcata* (the grey leafy lichen), *Melanelixia fuliginosa* (the dark crust) and *Candelariella vitellina* (yellow).

reasonable, but each particular example likely has its own peculiarities causing it to depart somewhat from this simplified story.

The ultimate examples of primary succession must be those that start on bare rock surfaces. Often lichens are some of the first organisms in the early stages of such successions (Fig. 146). These can be really common; for example, Seaward and Pentecost (2001) listed 69 species of lichen growing on bare rock within a 5 km radius of Malham Tarn Field Centre in the Yorkshire Dales. If you lie down on the famous limestone pavement at the top of Malham Cove and look at the apparently bare rock surface with a hand lens you will see it's almost entirely covered with a beautiful mosaic of various lichens, some actually living within the very top layer of the rock where enough light can penetrate to allow photosynthesis while the rock protects them from the feet of countless visitors. Admittedly most of these endolithic lichens are near impossible to identify by anyone who is not a specialist lichenologist! Over time other organisms, mosses for example, can start to colonise the rocks more easily once lichens have become established.

For secondary succession the main examples often used in textbooks, or undergraduate lectures, are ones of 'old-field succession'. This refers to the changes in vegetation seen in an agricultural field after farming use stops and 'natural' processes are left to take their course. Many of the examples come from eastern North America (Fig. 147), because during the eighteenth and nineteenth centuries lots of woodland was cleared for European-style agriculture, only to be later abandoned as people left the land to gravitate towards the growing cities of the east coast. After abandonment forest returned to these old fields, and because many major universities are now to be found in eastern North America these old-field successions have been well studied, although the studies have

FIG 147. Old-field succession in Ontario, Canada, photographed in 2004. This used to be the site of a farmstead which, like so many farms on the east coast of North America, was abandoned as people moved to the cities. At this site, at Queen's University Biological Station, the forest was cleared for agriculture around 1900, with a homestead built approximately where I was standing to take the photograph. The woodland has regenerated over the 50–60 years since the farm has been abandoned. The open vegetation in the foreground is prevented from succeeding to woodland by periodic disturbance from flooding and the activities of North American Beavers *Castor canadensis*, perhaps aided by deer grazing.

mainly had to rely on a chronosequence approach (e.g. Foster 1992). The two Rothamsted wildernesses described above are nice British examples of old-field succession, which have the advantage of being set up specifically as experiments and monitored over time rather than being interpreted as chronosequences at a later date. I will return to these experiments later in this chapter in the context of carbon storage.

MECHANISMS OF SUCCESSION

Two ideas that are often used to try and explain successional processes are competition (Chapter 6) and facilitation. However, other processes are important too, such as the dispersal of organisms to the site – if the initial disturbance has removed species from a location then if they are to return during succession they need to be able to recolonise. Some of these ideas look as if they should be, in principle, predictable – for example, the effects of competition. These are often referred to as 'deterministic' processes. Others, such as dispersal, may be more random (for example, chance events leading to seeds arriving at a particular location). The mechanisms of succession may be a complex mix of deterministic and more random processes, but the relative importance of deterministic and random processes is still unclear. Clearly successional processes are at least partly predictable: leave an open area, such as the sedge fen at Wicken, unmanaged and ungrazed and it will often turn into woodland, although the details of the type of woodland may be less predictable (think of the differences between the wooded Godwin Plot and the Godwin Triangle described above). As primary succession is the more dramatic process, given that it starts on bare ground, I will consider this first.

Organisms face two very obvious problems in the early stages of a primary succession. By definition, primary succession starts on bare ground without life, so organisms need to reach the site. In addition, the substrate they are colonising is often challenging, such as bare rock or blown sand. Because of the difficult conditions only a limited number of organisms are able to survive, assuming they can get there in the first place. The idea behind the concept of facilitation is that these early colonisers can alter conditions at a site, for example by adding nutrients and/or organic matter into the developing soils, which can make it easier for other organisms to colonise. In sand-dune systems the idea is that some of the early colonisers – such as Sand Couch or Marram – help stabilise the sand and add organic matter to the soil, making it more suitable for a wider range of species. Another classic example of primary succession is lichens colonising bare rock, which can increase the rate at which the rock weathers, thereby releasing nutrients that can potentially be used by other organisms (Zambell *et al.* 2012). In addition, the lichens are adding organic matter and providing a stable substrate that may be colonised by other organisms (note for example the small fragments of moss starting to colonise the rock surface amongst the lichens in Figure 146).

How important is facilitation? There is a general view that it is mainly important only in the earliest stages of primary succession (Keddy 2007). This is probably largely correct, but there are exceptions. Head back to Wicken Fen – or many other wetland nature reserves – and sit in a bird hide amongst the reedbeds (Fig. 148). Forget the birds for a moment and contemplate the plants. As Godwin and many other early ecologists realised, the reeds slowly reduce the water depth as peat forms from their dead leaves and stems. This eventually leads

FIG 148. Reedbed at Wicken Fen; without management this will tend to succeed to sedge fen.

to other fenland plants being able to colonise the now-drier peat surface. It's a pattern that doesn't just rely on chronosequences but, as described above, can be confirmed by analysis of peat cores. The reeds are allowing other plant species to colonise by reducing the water depth. Indeed, Keddy (2007) suggests that because their development is primarily driven by the accumulation of peat (composed of dead plant material) such peatlands are examples of systems where the idea of succession is particularly useful and easy to apply.

To visualise the role of competition in succession, think of the uncut Godwin Plot or the Godwin Triangle at Wicken – these have over the decades turned from open sedge fen to scrubby woodland, and in woodland the potential role of competition is very obvious. Stand in a deciduous woodland in summer and look up. Much of the sky is obscured by the leaves – and on the rare occasions where you can look down on the canopy then it's hard to see the ground because it's obscured by leaves and branches (Fig. 149). The trees are taking most of the light needed for photosynthesis, leaving much less for any plants growing on the woodland floor below. In the Godwin Plots some plant species are only found in the cut fen plots and are absent from the darker wooded plot – examples include Marsh Valerian *Valeriana dioica* and Early Marsh-orchid, which are mainly restricted to the plot cut every year. This idea of replacement of species during the course of succession also extends to woody plants, as shown by the

FIG 149. Woodland canopy viewed from above (from an observation tower) at Hickling Broad, Norfolk. In addition to illustrating the competition for light by woodland plants this observation tower is also one of the easiest places to see Purple Hairstreak butterflies *Neozephyrus quercus*, which spend most of their time high in the oak canopy, out of sight of the naturalist on the woodland floor.

replacement of Alder Buckthorn by Buckthorn in the Godwin Triangle. After the early stages of primary succession, when it is clearly important that a species must be able to disperse to and establish at the site, it seems likely that the key process at work is competition.

Climate is also important – so the old ideas of climatic climax contain a partial truth – and this is seen most clearly in the longer-term successions reconstructed from the analysis of preserved pollen in bogs and lakes. As the climate of Britain changed over thousands of years the vegetation changed too in response to these climatic shifts (Birks 2019). However, as Bastow Wilson and colleagues (2019) recently pointed out, despite succession being one of the oldest ideas in ecology we still know 'remarkably little' about the processes involved. As well as long-term records (e.g. from pollen analysis) and permanent plots (such as the Godwin Triangle), more actual long-term experiments are likely to be key to improving this situation. However, these take time and are difficult to run in a world which usually demands quick results.

SUCCESSION, CONSERVATION AND ENVIRONMENTAL RESTORATION

Consider primary successions; in the words of Tony Bradshaw, a pioneer of restoration ecology, writing in 1993, they 'demonstrate the creative role of nature in building up complete and complex ecosystems'. To appreciate this creative and transformative power, time-travel in your imagination to London sometime in the early 1940s. The Blitz has effectively constructed an experiment in ecological succession in the heart of the city; explosions and fire have created large areas of rubble which is now scattered a pinkish purple with the flowers of Rosebay Willowherb *Chamaenerion angustifolium* (placed in the genus *Epilobium* in the 1940s), with intermixed yellows from the flowers of various groundsels and ragworts. Some of the alternative English names of *C. angustifolium* commemorate this tendency to expand across destroyed cities – namely bombweed and fireweed. The feeling of the botanists who studied the flora of bomb-damaged London (particularly Edward Salisbury and Ted Lousley) was that Rosebay Willowherb did so well because of its liking for brightly lit habitats and the fact that it produces large numbers of wind-dispersed seeds which allow it to colonise newly created bare sites (Fitter 1945). So this wartime expansion of willowherb illustrates some of the points I made in the previous section about the importance of dispersal and competition in ecological succession. In a previous destruction of central London caused by the Great Fire of 1666 the aftermath was colonised by quantities of London-rocket *Sisymbrium irio*, but this didn't expand again following the Blitz – illustrating perhaps the importance of chance events in determining which species colonise and flourish during a succession.

In his book *Flora Britannica* Richard Mabey (1996) laments that he could find no record of how ordinary Londoners viewed the expansion of Rosebay Willowherb across their damaged city – did they see it as 'life triumphing over destruction' or just an untidy 'weedy invasion'? The response of artists and writers at the time crossed this divide; with for example Rose Macaulay's 1950 novel *The World My Wilderness* using this flowery primary succession to stand for barbarism, to be contrasted with the more orderly, neater, unbombed 'civilisation' (Matless 2016). By definition most of my readers will have an interest in natural history and are therefore more likely to view such successions in the way Tony Bradshaw did, as an expression of 'the creative role of nature'. A modern artistic use of the idea of succession in this more positive, restorative sense was the gold-medal-winning show garden by Howard and Dori Miller at the Royal Horticultural Society's Tatton Show in 2012 (Fig. 150). The garden 'World without

FIG 150. Ecological succession as artistic metaphor. The Millers' 'World without torture' gold-medal-winning show garden at the Royal Horticultural Society's Tatton Show in 2012. The top photo shows it under construction and the bottom one the finished garden. The photos show the second part of the successional sequence, running (right to left) from the arrival of the first saplings growing with some bomb-site rubble still visible (on the right) to 'climax' woodland with ferns (on the left).

torture' used the idea of a primary succession, starting on the rubble of a bomb site and wrapping its way around a fenced prison compound. The designers' idea was that the bombed area is quickly colonised by plants such as willowherbs and Butterfly-bush *Buddleja davidii*, and eventually tree saplings, until it becomes a stand of oak woodland with an understorey of ferns and mosses, in the designers' words 'showing how badly damaged land can heal itself over time'. The Millers were friends-of-a-friend, and came to talk to me about succession and the order you would expect plants to arrive. Indeed, it was the descriptions of the flora of blitzed London that inspired part of the eventual design.

The restoration of land changed by human activities can be seen away from the aftermath of war. Obvious examples are old quarries and sand/gravel pits, which, once abandoned, can revert to something greener, often becoming nature reserves. Two examples I know well are the Lancashire Wildlife Trust reserve at Mere Sands Wood and the woodlands surrounding the Centre for Alternative Technology near Machynlleth on the southern edge of Snowdonia. Mere Sands Wood (Fig. 151) used to be a commercial sand quarry until 1982,

FIG 151. Regenerating woodland at Mere Sands Wood, Lancashire – the site of former sand quarrying but now a nature reserve. Rhododendron *Rhododendron ponticum* agg. in flower. See also Figures 57 and 58 for more examples of habitats on the reserve.

but is now a mix of woodland and open water, with a smaller amount of more open terrestrial habitats. This range of habitats in a relatively small area made it a very useful site for both teaching and research when I was lecturing in Liverpool. The few older deciduous trees are mainly ones that were left around the edge of the site to screen the quarry from view. After active quarrying finished, birch woodland covered much of the site, and indeed one difficulty in managing this site is to prevent the fragments of grassland and heath succeeding to birch woodland. The Centre for Alternative Technology, where for several years I have been giving guest lectures to some of their Masters students, is situated in an old slate quarry (Fig. 152). Quarrying stopped in the 1950s and now much of the site has succeeded to woodland – rich in mosses due to the high rainfall in this part of Wales.

Sites like these show that in only 50 or so years succession can potentially go from highly disturbed industrial sites to woodland – though clearly at both of these sites it has helped that there has been an active interest in managing for conservation. Such natural changes are also obviously of interest for ideas of rewilding, an approach to conservation which tries to give priority to natural processes in undoing past human damage – a topic I return to in the final chapter. One of the reasons conservationists are interested in rewilding is as a possible response to climate change. In this context there is now a big interest in using trees to help reduce atmospheric carbon dioxide, so it's worth considering succession to woodland. To do so it's useful to return to the Rothamsted 'wildernesses', as these have been studied in some detail.

Over the course of about 120 years both Rothamsted wildernesses accumulated carbon as they succeeded from agricultural field to woodland.

FIG 152. The main quarry at the Centre for Alternative Technology, Machynlleth, photographed in 2019. This was a working slate quarry until the 1950s. In around 70 years woodland has returned to the hillside.

This in itself is hardly surprising, since it's what you would expect based on an understanding of photosynthesis as described in Chapter 3. However, the details are interesting and show differences in the amount of carbon accumulation between the Broadbalk and Geescroft wildernesses (Table 4). The Broadbalk Wilderness accumulated more carbon per unit area, in both soil and trees, than did the Geescroft Wilderness. Remember that Geescroft is the larger woodland while Broadbalk is more like a very thick tree-filled hedge. Paul Poulton and colleagues suggested that two factors, partly linked to the woodland size, were likely important in explaining why Broadbalk had sequestered more carbon than Geescroft. First, Broadbalk had more nitrogen available for plant growth, both from the soil at the site when woodland development started and probably also from more nitrogen entering the wood from nearby farms. As the smaller woodland (or very large hedge), Broadbalk has much more edge in relation to its size than Geescroft, probably allowing nitrogen from agricultural sources to blow into the wood more readily. Second, Broadbalk's smaller size, and so extensive edges, also allows more light to enter for photosynthesis. As well as these factors, the lower pH (higher acidity) of the Geescroft soil is probably due to a nineteenth-century history of more lime being added to Broadbalk when it was still being used for agriculture. Geescroft became much more acid during woodland development (its pH fell from an alkaline 7.1 in 1883 to an acidic 4.4 in 1999), and this has led to more leaf litter surviving on the woodland floor, at least in part because of reduced earthworm activity in the acid soil. In addition, in parts of Broadbalk succession was experimentally halted at grassland, either by pulling up tree saplings or in another area by grazing with sheep. These grassland soils have also tended to accumulate carbon but, because of the large amounts of carbon in the trees, the parts that were allowed to succeed to woodland ended up containing much more carbon than the grassland areas.

TABLE 4. Summary of carbon accumulation in the two Rothamsted wildernesses. Data from Poulton *et al.* (2003).

	Broadbalk Wilderness (wooded area only)	Geescroft Wilderness (all wooded)
Size (ha)	0.2	1.3
pH of surface soil in 1999	7.7 (slightly alkaline)	4.4 (acid)
Soil carbon accumulation over 120 years (t C ha^{-1} year^{-1})	0.54	0.38
Tree carbon accumulation over 120 years (t C ha^{-1} year^{-1})	2.85	1.62

Clearly ecological succession in much of Britain can lead to the development of woodland that, among many other things, increases the amount of carbon stored in both vegetation and soils. However, given the discussion in this chapter on some of the complexities of succession, which can lead to the details at different sites being different, it shouldn't be surprising that there are various complications. I will briefly discuss a couple of these issues. Although the two sites are close together, the rate at which trees colonised the Broadbalk and Geescroft wildernesses was different. At Broadbalk there was a dense thicket within 30 years of the start of the experiment, while at Geescroft after 30 years far fewer trees and shrubs were present. In this case the soils and past agricultural history of the two sites may have been involved in the different rates of woodland development. More generally, it's known that tree colonisation is often slower on more nutrient-rich agricultural soils, in part because grass grows better on such sites and this makes it hard for tree seedlings to compete (Harmer *et al.* 2001). Soil type can also be important in determining how much carbon is stored in soils under woodland. In their discussion of the Broadbalk and Geescroft results, Poulton *et al.* (2003) compared these sites with woodlands on more sandy soils. Both Rothamsted wildernesses accumulated carbon in their soils as the wood developed, but on sandy soils this may not happen, and pine woodlands may also accumulate much less carbon in the soils (although they do often have deep litter layers which contain carbon). The amount of carbon stored in grassland soils (such as grazed pastures) shows a similarly complex story, which makes it hard to generalise about the importance of this type of agriculture for carbon storage (Godfray *et al.* 2018). For example, in the grazed grassland plots at Broadbalk there was a lot of variation between soil samples in the amount of carbon in the soil, although the soils clearly stored more carbon than the arable land in the next-door Broadbalk Winter Wheat Experiment (Poulton *et al.* 2003). So, to generalise, these old-field successions tend to lead to an increase in carbon storage, but the details can be very site-specific.

Succession is also an important idea in British conservation, because many of our more valued habitats are not 'climax' but mid-successional and, left unmanaged, tend to succeed to other habitats – for example, the sedge fen at Wicken Fen or the hay meadows discussed in detail in Chapter 11 both require cutting to maintain that habitat long-term. Reedbeds (Fig. 148) are another good example; they are particularly associated with several species of rare birds but tend to turn into fen unless managed by cutting (Fig. 142).

Away from wetlands the use of livestock – mainly sheep, cattle or ponies – to manage succession for conservation management has become common. Usually they are used to prevent grasslands and heaths succeeding to scrub and woodland (Fig. 153). Even on wetlands such as Wicken Fen Konik ponies – which are considered tolerant of wet conditions – and Highland cattle are being used in grazing management, to manipulate successional processes. Often older rare breeds of livestock are used, as they are considered to be more tolerant of feeding on semi-natural vegetation, so these conservation grazing schemes have also played an important role in preserving older livestock breeds (Small 2015).

However, it's in the British uplands that the interactions between grazing and succession are particularly apparent. Historically high grazing pressure has helped maintain open grasslands in the British mountains, when the more natural ('climax') vegetation would often be woodland. With decreasing sheep grazing in the British uplands, shrub and trees are increasingly expanding. This is in general a good thing from an environmental perspective – more natural, providing habitats for a wider range of species, along with the potential for increased carbon sequestration and a role for trees in flood prevention. However, some species associated with open grassy habitats are now rare in the lowlands and potentially could be further threatened if there is widespread succession to woodland in the uplands where they can still be found. Examples include birds such as the Whinchat *Saxicola rubetra* and Skylark (Calladine *et al.* 2019).

FIG 153. Sheep grazing to halt successional processes as a conservation management tool near Corfe Castle, Dorset.

CONCLUDING REMARKS

Ecological succession was one of the earlier theoretical ideas in ecology (developed in the late nineteenth century). Despite occasional suggestions that the idea is old-fashioned, as Chang and Turner (2019) have pointed out, it still underpins many areas of ecology. Peter Moore (2001) used a nice comparison with the physics of light, and the question of whether light is a particle or a wave. He suggested that, as with light, there are multiple ways of looking at the 'universal but enigmatic process' of succession. The basics of ecological succession are easily summarised. Succession is the process by which biological communities reassemble following disturbance, be that natural (e.g. storm damage) or caused by human activity (e.g. mowing). It is conventionally subdivided into primary succession (that is, starting on bare ground) or secondary succession (starting on a site which contains some life – as in old-field succession).

Multiple mechanisms have been implicated in driving and explaining succession – such as competition, facilitation and dispersal limitation. The fact that some of these processes, especially dispersal, can involve large elements of chance helps explain why succession often seem less predictable than many earlier ecologists suggested. In a British context at many locations there tends to be a succession towards woodland (this is reasonably predictable), but the type of woodland, while being partly determined by climate, may also depend on chance processes and so differ between locations: hence Peter Moore's description of succession as somewhat 'enigmatic'. Therefore most ecologists now see succession as a dynamic process, rather than one heading towards a static climatically determined stable endpoint.

One problem with studying successions is that they take place over long periods of time. Many ecologists have tried to sidestep this problem by using space-for-time substitution – the chronosequence approach. However, this has the obvious difficulty that it assumes no other changes are affecting the system other than time since a particular disturbance event – and clearly this is at least partly unrealistic on a timescale of decades to centuries. Because of this, studies that have followed successional change over time (such as those at Wicken Fen and Rothamsted) or which use actual data from peat or lake sediment cores, are preferable to ones relying on chronosequences. However, such studies are relatively rare, and so we are often forced to use a space-for-time substitution despite the many caveats. Because many British habitats of conservation interest are mid-successional habitats (e.g. reedbeds, heathland and chalk grassland), the manipulation of succession is an important aspect of management on many nature reserves. So succession has become an important practical idea in conservation.

CHAPTER 10

Wytham Revisited:
Exploring the Ecological Niche

It is mid-morning in February, and as I arrive at Wytham Woods the sun is only just starting to burn through the mist. From the car park I hurry up the hill through the Lower Seeds pasture, looking to make photographic use of the mist before it vanishes for the day (Fig. 154). Photographs taken, I enter the woodland as the sky turns blue, following the path to the left along the Singing Way and into Marley Wood. With the sun now out, the male

FIG 154. Wytham Woods emerging from the mist, looking east from the fields of Lower Seeds.

Great Tits are starting to give their territorial 'teacher, teacher' calls, although plenty of these tits are still in their winter flocks. I stop to watch them foraging and notice that many of the Great Tits are feeding on the ground amongst the leaf litter. Feeding like this is textbook behaviour for these tits – indeed, the research on this was largely carried out here from the 1950s onwards. The reason why this work is so well known is that it provides a classic example of ecological niches and competitive exclusion, suggesting how different species competing for the same food can live together, and in what circumstance one species will exclude the other.

Marley Wood is an eastern extension of Wytham Woods running towards the River Thames and the city of Oxford (Fig. 155). The fact that it projects out into farmland and parkland means that it forms a reasonably distinct and

FIG 155. Marley Wood in early spring 2019. This part of Wytham has been used for studies of tits from the 1950s. Marley Wood forms a distinct unit, a promontory of woodland projecting out into surrounding parkland and farmland, and this was one of the reasons it was chosen as the location of the nest-box studies by David Lack's group. The shape of the woodland is, in part, because some of it was cleared in the nineteenth century by the fifth Earl of Abingdon to create a landscape of fashionable parkland views. In the 1980s there was a serious problem with deer, so Marley Wood was fenced in 2001 and has been largely deer-free for the last 20 years, allowing the regrowth of coppiced trees safe from browsing (Fisher *et al.* 2010).

isolated area of woodland, one of the reasons it was chosen as the location for an extensive study of tits in the 1950s, a study which is still running today (Perrins 1979). Parts of the wood are ancient coppice; that February morning the Hazel catkins were promising the imminent arrival of spring as I sat back against a tree trunk drinking from my flask of coffee and watching the foraging tits – Great Tits mainly amongst the leaf litter and Blue Tits amongst the catkins on the twigs above.

The ecological fame of the Wytham tits is very much intertwined with the career of David Lack (Perrins & Gosler 2010, Anderson 2013). At the end of the Second World War Lack was appointed as the new director of Oxford's Edward Grey Institute of Field Ornithology (EGI). Prior to war work on radar he had been a school biology teacher, and had carried out important studies on the Robins living around Dartington Hall School in Devon (see Chapter 6). His original idea on arriving at the EGI was to extend the Robin studies using Wytham Woods as his field site, as it had been given to the university three years before (see Chapter 3). Since that donation, the woods have been the site of a huge number of studies – it's a measure of how influential the ecological research here has been that it's the only site used to open two chapters in this book. However, Lack found that Robin nests proved rather difficult to find in extensive woodland, and this led to a rethink of his research plans.

Not long after arriving at the EGI, Lack visited the Netherlands and met Hans Kluyver (sometimes spelt Kluijver). Kluyver had been using nest boxes to study Great Tits, as their use of artificial boxes to breed in made it much easier to study them. Lack liked the idea, and so over the winter of 1946/47 Lack's student John Gibb installed the first 100 nest boxes in Marley Wood (Fig. 156). The start-up funds for this work came from a donation by Jane Wyndham-Lewis, the second wife of the writer J. B. Priestley. This wasn't quite the first use of nest boxes as a research tool in Britain. Around the same time Bruce Campbell took over a set of nest boxes at Nags Head in the Forest of Dean (now an RSPB nature reserve) which had been set up some years before in the hope of attracting birds that would help control insect pests of forestry. Kluyver's nest boxes had the same history, having also been first erected as a pest-control strategy, only later to become a research tool. This Forest of Dean study site – as with Wytham, it's still in use – focused on Pied Flycatchers *Ficedula hypoleuca*, but Campbell also collected data for Lack on tits (Campbell 1979). Much of the early work on tits around Wytham was carried out by Peter Hartley and two research students, John Gibb and Monica Betts (who later also worked at Nags Head following her doctorate – the first awarded to a woman for ornithology in Britain).

FIG 156. Nest box in Wytham Woods. This style of nest box has been in use since the early 1970s – the woodcrete (concrete and sawdust) box is woodpecker-proof, and the hanging attachment is intended to keep out Weasels *Mustela nivalis* (Perrins & Gosler 2010).

THE FEEDING ECOLOGY OF WYTHAM TITS

The tits of Wytham Woods posed an interesting question for mid-twentieth-century ecologists. There were five species found in the woods at that time (Great, Blue, Marsh, Willow and Coal), although Coal Tits and Willow Tits were relatively rare. All are rather similar birds living in the same habitat of deciduous woodland. One might expect such birds to compete with each other, and that the best competitor would outperform the others and come to dominate the woodland. So instead of five species of tits you would potentially end up with just one. An obvious thought is that perhaps the birds are not really competing, as they are using the woodland in different ways. In the case of obviously dissimilar birds this seems sensible – for example, look at the Chaffinch and Treecreeper in Figure 157, both caught for ringing on the edge of deciduous woodland by Malham Tarn Field Centre in Yorkshire. The very differently shaped bills of these two birds suggest that they feed on different things. This is indeed correct, with the Treecreeper eating mainly insects and spiders that it captures while foraging on tree trunks, while the Chaffinch is less specialised, feeding on a mix of seeds

FIG 157. The very different beak shapes of the Chaffinch *Fringilla coelebs* (left) and Treecreeper *Certhia familiaris* (right) strongly suggest that these birds eat different things, and so have distinct niches.

TABLE 5. Feeding locations (given as percentages; these are means of six months' worth of data) of the three commonest tit species in Marley Wood between November 1950 and April 1951, from John Gibb's doctoral studies (simplified from Lack 1971 and corrected for rounding errors using the original data in Gibb 1954). Note that the Great Tits are mainly feeding on the ground, while the Blue Tit specialises in feeding from thin twigs, and the Marsh Tit *Poecile palustris* prefers branches as a foraging location (twigs were defined as having a diameter less than 1.3 cm, branches were anything larger). There is obviously going to be variation over the six-month period: for example, ground feeding by Great Tits peaked in March and April, when 70 per cent or more of their time was spent foraging in the leaf litter (Gibb 1954). In addition to these three species of tits there were also much smaller numbers of Coal Tits *Periparus ater* and Willow Tits *Poecile montanus* in the study – the Willow Tit was lost from Wytham around 2002 (Perrins & Gosler 2010). These tits provide an additional example of organisms that have moved genus over the last few decades (see Box 1, Chapter 6) – all were placed in the genus *Parus* in Lack (1971).

	Great Tit	*Blue Tit*	*Marsh Tit*
Ground	50	7	16
Branches	16	8	30
Twigs, buds and leaves	8	37	21
Other	26	48	33

and invertebrates. But what about the Wytham tits? These five species not only look similar – confusingly similar in the case of Marsh and Willow Tits – but also have similar diets: a mix of invertebrates and seeds, with the latter particularly important over winter.

For his PhD work Gibb looked more closely at these tits and what they were actually doing. For example, he recorded where they were feeding over winter and found that although they appeared to be eating somewhat similar things they were concentrating their feeding in different layers of the woodland canopy (Table 5). He found that the lighter Blue Tits were spending much of their time gleaning food items off thin twigs and branches, while the heavier Great Tits were tending to feed on seeds from amongst the leaf litter, using their more powerful bill to open seeds (and it was the only one of the five species able to eat Hazel nuts; Fig. 158). These results are not specific to mid-twentieth-century Wytham Woods, and I remember replicating this work as an exercise when I was a Masters student – finding similar results from only a few hours of data collection on the winter feeding behaviour of Blue and Great Tits in woodlands on the bank of the River Wear in Durham.

FIG 158. Blue Tit *Cyanistes caeruleus* (left) and Great Tit *Parus major* (right) photographed at the same feeder to give a sense of size. Note the smaller size and less robust beak of the Blue Tit. Great Tits are typically about twice the weight of a Blue Tit, with a beak that is approximately 3 mm longer – 12.5 mm compared to 9.0 mm (Cramp & Perrins 1993).

These overwinter differences in feeding sites are not the only differences amongst the tits; for example, there were also feeding differences over summer and differences in nest sites used (Lack 1971). Other groups of birds show similar patterns. The alternative classic mid-twentieth-century example used in many introductory ecology books is work by Robert MacArthur (1958) on five very similar-looking warbler species in coniferous forests in Maine, USA. As with the Wytham tits, it appeared that they were using the same habitat in subtly different ways, feeding and nesting at different heights in the trees.

HABITATS AND NICHES

These birds are clearly using the woods in different ways. This is obvious when they are species as different as Treecreeper and Chaffinch, and more subtle for very similar tits or warblers. The word that probably most readily comes to mind for a naturalist thinking about these questions is 'habitat'. Although it's an often-used word, habitat is a rather informal concept denoting the place where an organism usually lives, such as oak woodland or chalk grassland. However, it's clear from the studies of Wytham tits, and many other organisms, that different species can use a single habitat in different ways. The term used to describe this is ecological niche, but this has been defined in a number of different ways over the last 100 years or so.

Early-twentieth-century ideas of the niche were not clearly differentiated from what we would now call habitat. The first influential concept of niche that was based on a fully worked-out definition was due to Charles Elton, in Oxford in the 1920s. This was before Oxford University owned Wytham Woods, and much of Elton's research at the time was on small mammals in Bagley Wood, a few kilometres south of Wytham and owned by St Johns College. In his book *Animal Ecology* (1927), Elton, who was a precocious 26 when it was published, defined niche as the role an animal is performing in a community – that is, what it is doing. He used an analogy with human communities and wrote that when an ecologist says 'there goes the Badger' they should be thinking about the Badger's role in the woodland community in exactly the same way that they would if they saw a human wearing a dog collar crossing the village green and said 'there goes the vicar' (Fig. 159). This view of ecological niches is usually referred to as the Eltonian niche, to distinguish it from other ways of defining niches that are now more widely used in ecology.

The idea of using analogies with human communities, and the range of different ways of earning your living that they can support, likely occurred to a number of people thinking about these problems in the early twentieth century. For example, in a short popular book on botany published in 1912 Marie Stopes

FIG 159. The Badger *Meles meles* was famously used by Elton (1927) in describing his idea of ecological niches; it's also an example of an ecosystem engineer, as discussed later in this chapter (see Fig. 168).

briefly used an analogy with the number of different trades in a town ('tanners and bakers and post men') to try and explain why multiple species of plants can survive in the same community. Although Stopes is now best known as a pioneer of sex education and contraception, at the time she was a leading expert on fossil plants. However, she didn't develop these ideas, and it's through Elton's more detailed discussions that this view of the niche became widely known (Wilkinson 2012). In summary, contrasting the terms habitat and niche, an oft-repeated way of comparing the idea of Eltonian niche with that of habitat is to say that the habitat is an organism's address, while its niche is its profession.

Although the Eltonian niche has been very influential, and still has its uses, since the mid-twentieth century the dominant concept of niche has been the one set out by Evelyn Hutchinson in 1957. Hutchinson's idea can be understood as building on the idea of limits of tolerance discussed in Chapter 2 (Fig. 16). Consider Snowdrops flowering in the February sunshine in Wytham Woods (Fig. 160). When growing in the wild in Britain their habitat is usually damp and often wooded – so they are plants of intermediate dampness, shunning both very dry and fully waterlogged sites. A graph showing their presence plotted against the water content of the soil would likely give an approximation of the bell-shaped curve shown in Figure 16, with their population peaking in damp, but not waterlogged, sites. However, soil moisture is not the only aspect of the environment of importance to Snowdrops; for example, it's not a plant you find high on mountains, although many mountainous sites may have the right moisture levels for the plant's survival. So you could potentially draw a graph with soil moisture on one axis and average air temperature on another and plot

FIG 160. Snowdrops *Galanthus nivalis* growing in Wytham Woods along the Singing Way. This common garden flower seems to have been introduced to Britain during the sixteenth century and is now well established in many woodlands and old grasslands, often with trees (such as graveyards). It's so widely found in the wild across many parts of the country that in the past some botanists have assumed it must be a native plant (Stace & Crawley 2015).

a shape on the graph that encloses all the conditions where both of these factors are such that they allow the plant to grow (e.g. the larger egg-shaped area in Fig. 161). Of course it's unlikely that only these two factors determine the correct conditions for Snowdrops – you could add a third factor on a new axis, giving a three-dimensional graph. However, it's also unlikely that only these aspects of the environment are involved, so you need to add more axes to the graph. This gives a graph that we can't draw, as it has too many dimensions, but you can imagine that such a graph might exist, and can mathematically describe it even though you can't visualise it (a mathematician would tend to talk about an n-dimensional graph, where n is a whole number that will vary from case to case).

What I have just described for Snowdrops is essentially Hutchinson's concept of the niche. In his words, 'an n-dimensional hypervolume is defined, every point in which corresponds to a state of the environment which would permit the species S_1 to exist indefinitely.' In school maths we are used to a volume being enclosed by a three-dimensional solid object, such as a sphere. However, mathematically you can do geometry in many more dimensions than just three, and it's such a multidimensional volume that Hutchinson was referring to by the term hypervolume. So the Hutchinsonian niche for an organism, such as a Snowdrop, is the range of conditions under which a population can exist, and it can be thought of as a hypervolume on a multidimensional graph.

In the Snowdrop example, what I have really been describing is the plant's fundamental niche – characterised by environmental factors such as soil moisture and chemistry, climate etc. However, for most organisms living in the wild other organisms – such as predators, parasites and competitors – can be very important. So a Hutchinsonian niche can take two forms. The fundamental niche is the range of conditions that the organism in question can theoretically survive and reproduce under in the absence of interspecific competition, predators and parasites. The presence of these other organisms, however, makes life more difficult and leads to the organism only being able to survive under a narrower range of conditions, called the realised niche (Fig. 161).

Economic analogies help make the point: a company may be able to survive under difficult economic circumstances, but if a couple of competitors also turn up then that may make the difference between survival and bankruptcy. Garden escapes, such as the Snowdrop in Britain, illustrate the difference between the fundamental and realised niches rather nicely. One, admittedly non-standard, way of viewing gardening is that it's all about allowing plants to be able to grow across the full range of their fundamental niche by removing competitors and pests via weeding, herbicides and pesticides (although readers of this book are probably more likely than the general population to limit the use of chemicals in any gardening they do). So with the help of active intervention by the gardener it is possible to get plants to grow in conditions that would be too marginal in the wild where they were having to contend with competitors, predators and parasites.

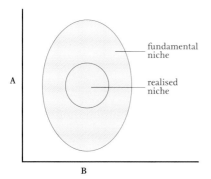

FIG 161. A simple two-dimensional representation of the idea of the Hutchinsonian niche (sometimes called the *n*-dimensional fried egg!). The graph shows two environmental variables (*A* and *B*) – for a plant, these could be soil moisture and pH. The pink shape shows the fundamental niche, the area where conditions of both *A* and *B* are suitable for the plant to grow and reproduce. The smaller circle shows the realised niche – the area it can occupy in the presence of competitors, parasites and predators.

In his 1957 paper Hutchinson made extensive use of mathematical notation in describing how he viewed niches, and his mathematically inspired definition may seem overly complicated. However, he included an insightful domestic analogy to justify his approach. He writes that in the day-to-day practice and application of ecological ideas such a 'logicomathematical' system isn't always necessary – just as it isn't always necessary to keep a vacuum cleaner in the middle of a room. However, 'When a lot of irrelevant litter has accumulated the machine must be brought out, used, and then put away.' The same follows for his mathematical approach to the niche: the rigour of this approach can be useful to tidy up confusions, but doesn't always need to be on show in every study. Mathematics can be very useful in science, not only for calculating particular results, but also in helping to get our ideas straight.[5] However, Hutchinson's ideas were developed in the mid-twentieth century, so an obvious question to ask is, how useful are they today?

Writing at the opening of the twenty-first century, in what was to become a widely cited paper, Ronald Pulliam (2000) described Hutchinson's niche concept as 'central to much of ecological reasoning and theory' as it 'provides a simple, although rigorous, approach to quantifying the niche'. However, as Pulliam went on to discuss, there are some problems when we think about the Hutchinsonian niche in the context of more recent ideas on source/sink populations and metapopulations (described in Chapter 8). It's now clear that on occasions we can find organisms apparently doing well in conditions where they are unable to maintain a self-sustaining population, as their population is being constantly topped up by individuals arriving from areas where conditions allow them to reproduce more successfully. So effectively we can find populations of organisms living in areas that fall outside their fundamental niche – as not all the requirements are available for long-term population persistence. In a similar way you may often find species absent from suitable sites. For example, in a woodland Primrose tends to be found in places where gaps in the tree canopy allow more light to reach the forest floor (Fig. 162). When a new gap forms it may take time for Primroses to colonise – either from seeds or by expansion of a few plants that had managed to survive in the shade (albeit under conditions not suitable for their reproduction). So Primroses within a wood can be absent from sites with the correct conditions for growth and reproduction, and also sometimes present at sites where they could not reproduce (Valverde & Silvertown 1997). So if we

---

5 For the mathematically inclined, Hutchinson's 1957 paper made use of set theory, a common approach for mathematicians trying to be rigorous about their ideas. Indeed, most mathematicians would view 'set' as an even more fundamental concept than 'number' (Stewart & Tall 2015).

FIG 162. Primrose *Primula vulgaris* flowering in a sunny patch in a recently coppiced part of Bunny Old Wood, Nottinghamshire.

want to understand why a particular species is found at (or absent from) a site we need to consider not only the fundamental and realised niches but dispersal too.

GENERALISTS, SPECIALISTS AND NICHE BREADTH

Some species seem to have much larger Hutchinsonian niches than others. My favourite example is the Raven (Fig. 163). In Britain, I have watched Ravens on cold rain-swept mountains but also in the lowlands and in some cities too; abroad, I have watched them high in the Alps, around urban-fringe rubbish tips in a hot dry Mediterranean summer, and in unpleasantly hot conditions in the arid mountains surrounding Death Valley in the USA. Contrast this with another of our corvids, the Chough (Fig. 164). In Britain this is a much rarer bird, associated with clifftops in the west of the country and also occasionally in the mountains. In the words of Cramp and Perrins (1994), the Chough's 'restricted and fragmented distribution suggests strictly specialised needs, as does widespread susceptibility to declines in numbers'. A national survey of Choughs in 2014 estimated a population of 433 breeding pairs – mainly in Wales and the

FIG 163. A Raven *Corvus corax* in Cwm Idwal, Snowdonia. This species is widely distributed across the northern hemisphere and found in a very wide range of habitats, from forests to deserts to mountains, as in this Welsh example.

FIG 164. The Chough *Pyrrhocorax pyrrhocorax* is a much more specialised crow species, in Britain now mainly restricted to the western coast. This colour-ringed individual was photographed on sea cliffs on Anglesey, which had around 9 per cent of all the breeding pairs in the UK in 2014 (Hayhow *et al.* 2018).

Isle of Man. The vegetation seems key to Chough success, with a short sward allowing access to invertebrate food in the soil and in animal dung. Grazing level appears crucial, with either too little or too much grazing creating the wrong conditions for these specialist birds (Hayhow *et al.* 2018). This raises an obvious question. Surely it's better to be a generalist like the Raven, able to use a wide range of resources and habitats – so why do specialists exist? The likely answer is based on the ideas of trade-offs introduced in Chapter 1. In the right conditions a specialist can do very well – the Chough's long red bill is very good at accessing soil invertebrates in the right conditions – but optimising the bill for these conditions makes it less good for use in other ways.

The idea that some organisms are specialists, some generalists, and presumably some part way between these two extremes, makes sense. However, it would be good if we could make this idea more quantitative, so that we could score different organisms on their level of 'specialism' and so study these ideas in a more rigorous way. One commonly used approach to doing this is Levins' niche breadth, devised by Richard Levins in the 1960s. This uses the number of resources used by an organism to create a quantitative score describing the size (breadth) of its niche (Southwood & Henderson 2000). So in the corvid example the Raven would have a larger niche breadth than the Chough.

There are two different approaches to calculating niche breadth, both of which could be applied to Ravens and Choughs. One is the approach I took when I first compared these two species at the start of this section, namely identifying the number of different habitats used by each. A better approach might be to quantify the range of food items eaten by each species (either by direct observation or perhaps looking at stomach contents). For many organisms the first approach, using range of habitats, is often more practical – although it comes with problems. If we were interested in comparing the niche breadths of different herbivorous insects we might sample different vegetations with a sweep net and look at which species were only found associated with a small number of plants, and which were more widely distributed. This also illustrates the problem: we might record a species from a vegetation that it doesn't really use if it's a stray individual just passing through. However, for many organisms, such as bacteria, we know so little of their natural history that we have little choice but to use this habitat-sampling approach.

COMPETITIVE EXCLUSION

We can use Hutchinson's approach to niches to think about the Wytham tit data in Table 5 and use it to introduce the important idea of competitive exclusion.

The Wytham tits, and in North America Robert MacArthur's warblers, feature prominently in many introductory ecology courses because not only do they nicely illustrate the idea of niches, but they are also relevant to the concept of competitive exclusion. The basic idea behind the competitive exclusion principle is that two species cannot survive at the same site indefinitely if they are both limited by the same resource – one of the species will be better at competing for that resource, and even if it's only a slightly better competitor, over time that should lead to it outcompeting the other species. The idea can be phrased even more succinctly as 'complete competitors cannot coexist' (Hardin 1960). This would suggest that the reason why several rather similar species of tits can be found in many British woodlands – including Wytham – is that they are using the habitat in different ways and so are not 'complete competitors'.

In describing Joseph Connell's 1950s experiments on the Isle of Cumbrae in Chapter 6, I made the point that one of the things that was important about them was that experiments on competition carried out in the field were rare at the time, although there were some well-known laboratory experiments (or garden experiments, in the case of Arthur Tansley's early-twentieth-century studies of competition between species of bedstraws). Some of the most famous of these early laboratory experiments were carried out by the Russian biologist Georgii Gause, and used protozoa, especially ciliates in the genus *Paramecium*. Species in this genus are widespread in aquatic habitats, and can be particularly common in small ponds with organic muds. Gause (1934) used laboratory cultures of *Paramecium*, with other microbes as their food source. He combined the data from these experiments with mathematical analysis to investigate a number of key theoretical ideas in early-twentieth-century ecology, such as logistic growth, models of the interaction of predators and prey, and competitive exclusion. The interplay of laboratory experiments and attempts to construct mathematical models of what was going on is one of the great strengths of Gause's work; indeed, the modern reprint of his classic 1934 book – sitting on my desk as I write this section – is rightly described as 'a classic of mathematical biology and ecology' on its cover.

There is an interesting story attached to these famous experiments, which I have already briefly mentioned in the discussion of logistic growth in Chapter 8. In many of his experiments – especially those focused on questions of population ecology such as logistic growth – he used yeast as a food for two species of *Paramecium*. However, we now know that while these species of *Paramecium* can ingest yeast they can't actually live on a diet of yeast. As I have already suggested, the likely explanation is that the yeast cultures were infected by bacteria and the *Paramecium* were feeding on these. However, in the key competitive exclusion experiments – described below – Gause actually used the bacterium *Pseudomonas aeruginosa* (known at the time as *Bacillus pyocyaneus*) as food rather than yeast. Again it wouldn't be surprising if these cultures contained a wider range of bacteria than Gause intended. Several studies have found that *Paramecium* tends to grow better if fed a range of different species of bacteria, although it does appear to do alright on some strains of *P. aeruginosa* (Jones 1974). However, many of these studies of *Paramecium* feeding on bacteria are from the early twentieth century, so I do wonder about levels of contamination in their supposedly single bacterial species cultures. These experiments provide a nice illustration of the role of luck in science; although things were more complex than Gause realised, he was lucky, and it appears that what he thought were high food cultures were indeed that – although in many cases the food probably wasn't the species of microbe he thought it was.

In the experiments most directly relevant to this chapter, Gause used two different species of *Paramecium,* namely *P. caudatum* and *P. aurelia.* When populations of each of these were grown separately with bacteria as food the populations grew quickly at first before levelling off at the carrying capacity (providing nice examples of logistic growth). However, when the two species of protists were grown together the result was somewhat different. *P. aurelia* still showed logistic growth, whereas the population of *P. caudatum,* although it started by increasing, later declined to near extinction. The interpretation was that *P. aurelia* was the better competitor and likely reduced the density of bacteria to a point where *P. caudatum* was unable to persist, and so was ultimately excluded from the culture – hence 'competitive exclusion'. Because of these experiments the idea that I am calling competitive exclusion is sometimes also referred to as 'Gause's principle'. This usage seems to be largely due to David Lack, who was very interested in Gause's ideas when it came to interpreting his own observations on the finches of the Galapagos Islands. However, many other people had made similar observations or theoretical suggestions prior to Gause. In Garrett Hardin's (1960) nice analogy, if the idea was born with Gause it was certainly conceived quite a bit earlier.

There are, however, difficulties with the idea of competitive exclusion. One obvious one is that while the idea states that two species cannot coexist indefinitely if they have exactly the same requirements – that is, share the same Hutchinsonian niche – it's not clear how to test this idea with real data. If, for example, you find two species of tits coexisting and apparently sharing the same niche, have you shown that the idea doesn't work in this case, or have you perhaps just failed to measure some important aspect of their environment which the two species use in different ways? Also, there are some ecological systems where at first sight it seems extremely unlikely that competitive exclusion could be at work. A striking example is provided by phytoplankton in lakes or the ocean, and this issue is often called 'the paradox of the plankton' after a famous article by Evelyn Hutchinson (1961) that drew attention to the difficulties plankton raise for the idea of competitive exclusion. Hutchinson's problem was how to explain the fact that there are lots of species of photosynthetic plankton in most water bodies. For example, in a study at Rostherne Mere in Cheshire (Fig. 165), over the summer of 2000, species from six different major groups (phyla) of phytoplankton were found in water samples from the mere (Fisher *et al.* 2009). They all have apparently similar requirements – such as light and nutrients – so why don't just a small number of species outcompete all the others?

There are a number of potential answers to this paradox. One that Hutchinson himself discussed in some detail is that the environment is changing at a rate which means that the community can never settle down into equilibrium conditions such that one species can successfully outcompete another. For example, if one species does a bit better at lower water temperature it probably does less well as the summer progresses and the water warms up. To use Rostherne Mere again as an example, in early May 2010 the surface water temperature was around 12 °C, but later that summer, around the start of July, it

FIG 165. Rostherne Mere in Cheshire (see Fig. 71 for another view of the Mere and further details of the site).

peaked at 21 °C (Reynolds 2019). So by the time a colder-water species might have started to outcompete a warmer-water one, the water temperature has changed to suit the warmer-water one, and before this can outcompete the colder-water species the winter is arriving and the water is getting colder again. In addition, at the sizes relevant to smaller plankton, such as bacteria, it's very difficult to visualise how they are experiencing the environment. What looks like a well-mixed body of water to a human swimmer might appear very different at the very small scale. The total weirdness of the physics of water at this scale, at least to a large animal like a human, is illustrated by the fact that a bacterium trying to swim like a fish by wagging its tail to and fro would fail to make any progress at all. Because of this, bacteria have to take very different approaches to propelling themselves through water (Thorne & Blandford 2017). The physics experienced by microorganisms is very strange, and it's hard to understand exactly how they experience their environment.

COMPETITION BETWEEN DISTANTLY RELATED ORGANISMS?

It's common to suggest that the strongest competition is between closely related species (or individuals of the same species), as they have very similar requirements – and I have suggested so already in this book. However, there can also be strong competition between distantly related organisms. Consider a rocky shore, a classic habitat for student field trips and rock-pooling naturalists (Fig. 166). Many very distantly related organisms can be competing for space on the rocks, from brown and green seaweeds (two very distantly related groups of algae – see Chapter 4), to lichens and animals such as barnacles and mussels.

Away from the shore, an intriguing possibility is that there may be competition between organisms as different as a Blackbird and a bacterium concerning who gets to consume fruit. Consider a windfall apple lying beneath an apple tree (Fig. 167). This is food for a wide range of microorganisms, and indeed this is what is happening as the apple rots: it is being broken down by a collection of fungi, bacteria and other microbes. However, the apple is also potential food for animals such as thrushes, and an apple eaten by a bird is not available for use by microbes, and vice versa, so there is a potential for competition.

Back in the 1970s the American ecologist Dan Janzen paid 95 cents for an avocado that turned out to be rotten. This set him thinking about mouldy fruit, and he wondered if some of the chemicals given off by the microbes in rotten fruit (or in an animal carcass) could be produced to stop animals eating them.

FIG 166. A rocky shore at low tide – a habitat where naturalists, biology students and holiday makers all compete for the best rock pools. In this case it's the shore at Rhosneigr on Anglesey.

FIG 167. A windfall apple being colonised by fungi and other microbes. Is there competition for the apple as a resource between microbes and large animals such as thrushes?

In the technical paper he wrote on the idea (Janzen 1977) he credits the purchase of the avocado as the initial source of his ideas. Some birds certainly share our preference for fresh rather than rotten fruit, so the idea looks plausible – but as Janzen himself pointed out, there is a problem. Clearly each individual microbe can only produce a small amount of toxin, or alcohol, to deter animals from eating the fruit – so to be effective lots of individuals have to work in unison. Effectively the problem is very similar to that of taxation in a human society. Everyone benefits from the services the state provides, but if one person decides to cheat by not paying tax then that individual still benefits from all these services while saving money. The logical conclusion is that everyone should decide not to pay, and should rely on services provided from the taxes of everybody else. Obviously the system then collapses, and the hospitals and schools close through lack of funds – which is why the paying of taxes is enforced by governments, to preserve these public goods. The same idea applies to microbes on an apple producing chemicals to keep the thrushes at bay. If one microbe stops producing these chemicals it saves on resources and still benefits from the chemicals produced by everybody else; so the system should be unstable, breaking down as all the microbes start to cheat and rely on other microbes' chemical defences. Could there be situations where the production of chemicals to deter vertebrates could evolve, or is this no more than an intriguing idea that doesn't actually work in the wild?

On way to investigate questions like this is to use mathematical models to investigate what, in principle, might work. There have been two attempts to model Janzen's ideas mathematically. I was involved with both of them, and the interesting thing is that these two different models arrived at different conclusions (Sherratt *et al.* 2006, Ruxton *et al.* 2014). The details of the maths are not needed here – except to point out that our second 2014 model was very explicitly a metapopulation model, based on earlier work by Sean Nee and Robert May (for metapopulations, see Chapter 8). Our first model suggested that – assuming making these chemicals had a cost for the microbes – it was not possible for the chemical production to have evolved as an adaptation for deterring vertebrates, for the reason suggested by Janzen (because of the problem of cheats taking over the system).

Our second go at modelling this idea, eight years later, found that under some conditions these chemicals could, in principle, evolve for the reasons Janzen suggested. The key thing, which took us some time to work out, was what were the biological implications of this difference, i.e. the biologically relevant differences between the two models? The answer was the way in which the two mathematical models described dispersal of microbes between the apples (or dead animals in the case of carrion). In our 2006 model all 'apples' tended to be colonised by all types of microbes, and the ones that didn't have the cost of producing these chemicals grew better and outcompeted the other microbes. In the later model dispersal between 'apples' was more restricted, leading to some being colonised only by chemical-producing microbes (and these survived for longer, as their apples were less likely to be eaten by birds). This is really a special case of a widely known result in metapopulation theory – that it's possible for a weaker competitor to survive long term if it's better at dispersal and/or less vulnerable to the extinction of its patches (Hanski 1983).

Our two models tell us something about Janzen's ideas in particular, but also more generally illustrate one of the ways in with you can apply maths to ecological theory. The models cannot show that Janzen's idea is correct, but they can show that in theory it could (or couldn't) work. Beyond that, they identify rates of dispersal as a particularly important aspect, so potentially providing a guide for future experiments addressing the idea. There are, of course, other reasons why the microbes might be producing these chemicals. For example, they may just be waste products of the microbes' metabolism, and/or produced not to deter vertebrates (Janzen's idea) but as part of competition between different microbes colonising a windfall apple or cadaver. In a complex real world, away from the simplifications of mathematical models, it's quite possible that all of these explanations are playing a role. However, the next time you watch thrushes feeding on windfall apples, consider the possibility that you might be watching competition between organisms as distantly related as Blackbirds and bacteria.

STATE OF THE ART – RECENT IDEAS ON ECOLOGICAL NICHES

This chapter so far has mainly described classic ideas on the ecological niche, mainly developed during the middle of the twentieth century. Almost every idea I have discussed so far could have featured in a lecture that I might have given when I started teaching university classes in the early 1990s. However, over the last 30 years or so a variety of new ideas have been introduced, and these are the subject of the final part of this chapter. In 1994 Clive Jones, John Lawton and Moshe Shachak published a research paper in the academic journal *Oikos* entitled 'Organisms as ecosystem engineers'. This has become a very influential paper, and it's extremely unusual for an important paper in theoretical ecology published in recent decades in that it's all based on observational natural history, without any use of mathematics at all (many of Dan Janzen's papers also have this same unusual aspect to them).

The key point that Clive Jones and colleagues made was that some organisms are particularly important because they alter the environment in ways that affect other species – in either a positive or a negative way. That is, they emphasised 'the role that many organisms play in the creation, modification and maintenance of habitats'. One example they used in their paper was the sett-digging activity of Badgers – the setts can provide shelter not only for the Badgers but also for other animals such as Foxes and many invertebrates (Fig. 168). Viewed one way,

FIG 168. The Badger *Meles meles* as an ecosystem engineer. Badger setts (above) provide a home not only for Badgers but also for a range of other organisms such as Foxes *Vulpes vulpes* (below).

this appears totally trivial, for most naturalists could list many examples of cases where one species affects the habitat and niches of many others. However, it's often the case that we tend to overlook and downplay the obvious, and what was important about this *Oikos* paper was that it drew attention to the importance of ecosystem engineering as a general phenomenon. Indeed, bestowing the name 'ecosystem engineers' was an important step; as Stephen Jay Gould (2002, p. 875) commented in another context, 'phenomena without names, and without theories marking them as worthy of notice, will probably not be recognized at all'.

Various examples of ecosystem engineering have already been discussed in this book, albeit without necessarily labelling them as such – for example, the role of Beavers in modifying the environment by building dams (Fig. 26), or the idea of facilitation in a succession as described in Chapter 9 (Fig. 169). However, there is

FIG 169. Primary succession on a very small scale: the mosses on this fence post in Cwm Idwal are engineering the environment, making it suitable for the small Heather *Calluna vulgaris* seedling. The mosses are acting as autogenic ecosystem engineers, modifying the environment by their physical structure (for example, by retaining water).

a big difference between the actual engineering work carried out by Beavers, and the way that the growth of one type of plant improves the conditions for others, so Jones, Lawton and Shachak differentiated between two different kinds of ecosystem engineers, allogenic and autogenic. Allogenic engineers (e.g. Beavers, woodpeckers, humans) are species that engineer in the human meaning of the term, transforming things from one physical state to another. Autogenic engineers (e.g. the trees in Wytham Woods, or the bog mosses discussed below) change the environment via their own physical structure – for example, trees shading the woodland floor.

The idea of ecosystem engineers overlaps somewhat with an older idea, developed in the 1960s, of 'keystone species'. These are species that, while often not that common, have substantial effects on the environment of other organisms. Some, like Beavers, are ecosystem engineers, while others, such as predators, are not. In addition, some ecosystem engineers may have only minor effects on the environment, and so wouldn't be classed as keystone species – think of the footprints left by a large mammal, which may provide a suitable site for the germination of some plant species. So ecosystem engineers may sometimes be keystone species but not always, and the concepts are rather different (Jones *et al.* 1997).

The bog mosses *Sphagnum* spp. are classic examples of autogenic ecosystem engineers, and indeed they featured as an example in the original *Oikos* paper by Jones and colleagues. Britain and Ireland have around 34 species of bog moss, ranging in colour from vivid greens to crimson reds (Fig. 170). They are key plants in many peatlands, not only growing in these often nutrient-poor peaty habitats, but helping to actively create the habitats. The niche of a typical bog moss tends to be both very wet and also acidic – conditions in part engineered by the mosses themselves. The mosses can absorb large quantities of water, and in the past they have even been used on a large scale as wound dressings, as dry moss can absorb 3–4 times as much water as a cottonwool dressing (Daniels 1989). Part of the explanation for this ability is their leaf structure, which has large dead empty cells (hyaline cells) that can fill with water (Fig. 171). In addition, the spaces between the leaves and stems in a densely packed carpet of bog moss act as a wick for water movement. So the presence of these mosses tends to make the habitat wetter.

The bog mosses also make the habitat more acidic. They are adapted to growing in nutrient-poor conditions, and the mechanism by which they acquire the limited available nutrients involves giving off positively charged hydrogen ions, leaving behind negatively charged compounds on the moss surface which can bind with many nutrients. Readers who remember a bit of chemistry may recollect that acidity is related to hydrogen ion concentration (in fact, pH is the negative log of the hydrogen ion concentration), so these processes make the

FIG 170. Multiple species of bog mosses in Abernethy Forest in the Cairngorms. The bright red bog moss is *Sphagnum capillifolium* subsp. *rubellum* (sometimes considered a full species rather than a subspecies of *S. capillifolium*). In more shaded conditions they tend to be less brightly coloured. There are around 20 peatlands (mainly nutrient-poor ombrotrophic sites) scattered through the forest, with *S. capillifolium* a common bog moss on the majority of these sites (McHaffie *et al.* 2009).

habitat more acidic. Other types of moss (brown mosses) can do this too, but the bog mosses are particularly good at acidifying their environment in this way (Rydin & Jeglum 2013). So the wet and acidic niche required by many species of bog moss is in part engineered by their own presence.

This simple description of the basic idea of ecosystem engineers – that they affect the environment via their physical presence and/or behaviour – glosses over important details. By altering their environment they potentially affect their own success in a positive or negative way. For example, the bog mosses are tending to make the environment more suitable for themselves, but other mosses in a primary succession may be making the conditions more suitable for larger plants to colonise and potentially outcompete the mosses (Fig. 169). So there are feedbacks between the engineers and their environment, with each affecting the other (Jones *et al.* 2010). These feedbacks can be considered over an evolutionary timescale as well as over the (typically) shorter ecological timescale I have just discussed. These two approaches are combined in a related idea called niche construction (Odling-Smee *et al.* 2003). This concept has its origin in population

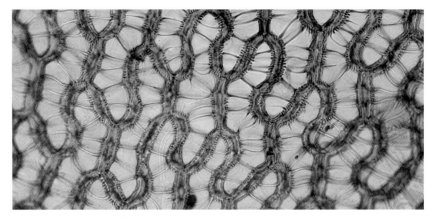

FIG 171. Hyaline cells of bog moss viewed under a microscope at x400 magnification (each cell is approximately 0.03–0.04 mm wide). These dead empty cells can hold large amounts of water. The species photographed is *Sphagnum austinii* (previously named *S. imbricatum*), now rare in much of Britain although it used to be much more widely distributed. The specimen in the photograph came from a peat core from Holcroft Moss in Cheshire; in Britain the species is now mainly restricted to Scotland, with a scatter of sites in north Wales and northwest England (it is no longer found in Cheshire). The hyaline cells of this species are particularly distinctive, being lined with comb-like structures (comb fibrils) that can be seen in this photograph.

genetics rather than ecology, and suggests that, rather than organisms just adapting to niches set by the environment, over time their own actions can help create the niches that they occupy – think of bog mosses creating peat bogs and ongoing adaptations in the mosses to adapt to these boggy conditions.

The background population genetic theory for niche construction is to a large extent based on the ideas of Richard Lewontin. During the 1960s Lewontin, along with a group of influential young scientists such as Robert MacArthur, Richard Levins and Edward O. Wilson, tried to bring the rigour of simple mathematical models to ecology. Traditionally evolution was thought of as operating over longer timescales than ecological processes – so the two could be considered separately. However, Lewontin, Levins and colleagues realised that this is too much of a simplification. For example, succession (an ecological process) can take decades, if not centuries, while some forms of evolution (e.g. pesticide or antibiotic resistance) can happen over a few years or less. So they argued that ecology and evolution need to be considered together (Odenbaugh 2013). Niche construction grew out of this approach, and, as noted by its proponents, there are similarities between it and some aspects of the concept of Gaia as proposed by James Lovelock and Lynn Margulis (discussed in Chapter 5).

These recent ideas on ecosystem engineering and niche construction envisage much more of a two-way interaction between life and the environment than was typical of the mid-twentieth century when Hutchinson was developing the basics of modern ideas of the ecological niche. One of Hutchinson's books was entitled *The Ecological Theatre and the Evolutionary Play* (1965), the idea being that the environment is the background against which organisms develop adaptations, so that ecology provides the setting for the ongoing drama of evolution. Many ecologists now see a much more intertwined connection between ecology and evolution, with each affecting the other. The actors are doubling as stage hands, moving the furniture around as the play progresses.

CONCLUDING REMARKS

Returning to Wytham Woods, we can add detail to the simple description of the habitat as largely deciduous woodland. The woodland potentially contains large numbers of Hutchinsonian niches, each of which could in principle be described in relation to a large number of variables – such as moisture, temperature, amount of light and so on. The large number of niches is one of the reasons why so many species can survive in the wood (although the paradox of the plankton warns us that this is unlikely to be the only explanation), so the concept of niche is important in conservation, where we are often trying to maintain this biodiversity – a topic that I return to in Chapter 12. In this niche-based approach to understanding the biodiversity of the woodland, competition is important both in reducing the niche size available to a given organism (remember that the realised niche, with competitors present, tends to be smaller than the fundamental niche), and potentially in determining the number of species able to survive in the wood long-term through the process of competitive exclusion – although again the paradox of the plankton cautions us against accepting this as the only explanation.

Some of the species in the wood are also of particular importance because they modify the conditions for other species, and these are ecosystem engineers. Obvious examples are the tree species, which greatly alter the conditions on the woodland floor through shading and other processes. The well-studied Badger population in the woods is also engineering the environment of other organisms through sett construction, while one of the Badgers' main sources of food, the earthworms, are also engineering the soil structure. So the concept of the ecological niche has become a little more complex than its mid-twentieth-century version, expanded to encompass two-way feedbacks between organisms and their environment.

It's clear that the concept of niches is an important one in ecology, but long-standing problems, such as the paradox of the plankton, suggest that there may be limits to its ability to explain some patterns we see in the wild. Consider a woodland like Wytham: the idea that many of the species in the wood utilise different niches makes obvious sense in many cases, for Snowdrops, Badgers, Great Tits and Oak trees all require different things from their environment. However, as pointed out earlier in this chapter, applying ideas of source and sink populations suggests the possibility that some of the species you find in the wood may not be able to survive there long-term as self-sustaining populations because the wood does not provide all that they need for this. These species are in the wood because of dispersal from sites elsewhere which better match their requirements (or as hangovers from a past when conditions suited them better).

Over the last 20 years Stephen Hubbell (2001) has argued that at the scale of whole forests or lakes the effects of competition and niches may not be important, and that you can explain the numbers of species found there by ignoring details of their requirements and just treating all species as having effectively the same requirements and assuming that births, deaths and dispersal are effectively random. This is the 'neutral theory of biodiversity', and it has been controversial because it's obvious to any naturalist that species are not the same and that they do have different requirements of the environment. However, Hubbell's claim is that at large enough scales these details can safely be ignored. One analogy is with the kinetic theory of gases in physics – here, rather than trying to follow the detailed behaviour of every atom, physicists instead use a statistical analysis of the behaviour of the whole population of atoms to understand what's going on. As Willard Gibbs – one of the key figures in the development of this area of physics in the nineteenth century – wrote, 'One of the principal objects of theoretical research is to find the point of view from which the subject appears in the greatest simplicity' (Thorne & Blandford 2017). The neutral theory has a similar philosophy of seeking simplicity in broad statistical patterns while ignoring the detail. This has become a large research area, underpinned by complex mathematical techniques that attempt to identify neutral versus niche-based patterns in ecological data sets. There is a growing consensus that in many cases both niche-based and neutral processes may be at work in the same ecological community – the relative importance of each depending on the details of the community under study (May *et al.* 2007, Huneman 2018). Regardless of the details, however, niche concepts are still central to many ecological explanations, and are still evolving.

CHAPTER 11

Park Grass and the Hay Meadow Conundrum

Harpenden, just north of London. Our coach turns off the main road and through a cluster of modern buildings – all brick, metal and glass, looking like many a small academic campus anywhere from North America to China, with only the use of brick making it in any way individual. This is Rothamsted Research, and despite the modern architecture it's the oldest agricultural research station in the world. Our coach continues past this modern campus and through a small patch of agricultural countryside hemmed in by urban development and golf courses, and stops by an old manor house – the architecture of a bygone scientific era, as the first on-site laboratory was set up here in the nineteenth century. Once the occupants of our coach have disembarked and assembled, it's just a short walk through a small fragment of woodland to our destination: a long thin rectangular hay meadow, bounded at one end by Redbourne Lane (the B487).

But what a meadow! Seen from the air (or on Google Earth) it's a patchwork of multiple different squares and rectangles – each a different experimental treatment. This is the Park Grass Experiment, and it is often called the longest-running ecological experiment anywhere in the world. That's why a coach load of ecologists have come on an excursion to view this famous site. It is September 2005, and the coach has brought some of the delegates from the annual meeting of the British Ecological Society, which is being held at the University of Hertfordshire that year. Almost 90 years earlier, in the midst of the First World War and less than two months before the carnage of Passchendaele, one of the first outings of the Society had also been to Rothamsted. Writing many years later, Edward Salisbury (1964a) reminisced that:

after I became Hon. Secretary of the Society in 1916 the first field excursion that I arranged was, by general desire, to Rothamsted Experimental Station and in particular to study the classic grass experiments.

Salisbury and his group spent the afternoon examining the Park Grass plots before being entertained to tea in the laboratory. We also had just an afternoon to look around, but without the tea party.

Our guide in 2005 was Jonathan Silvertown (Fig. 172), who started studying the ecological implications of this experiment as a PhD student in the 1970s. Although he has since worked on a wide range of topics in plant ecology he has maintained an interest in the experiment, and much of my discussion of the Park Grass Experiment draws on a 2006 review paper he wrote with several colleagues in the *Journal of Ecology* (Silvertown *et al.* 2006).

What, then, does the Park Grass Experiment tell us about the way the world works? The answer is rather a lot – but in this chapter I concentrate on what it tells us about why some kinds of vegetation contain lots of plant species, while others are rather species-poor, and then put this in the context of other hay meadows and the wider world. However, it has also contributed to our

FIG 172. Jonathan Silvertown explaining the Park Grass Experiment to visiting ecologists in September 2005. The white posts mark the various plots within the experiment.

understanding of small-scale evolutionary changes in plants – as it's been running long enough for such changes to start to become apparent. It also provides an important example of long-term stability in grassland vegetation – it's been running long enough for the lack of changes in the control plots to be important too. How did such a famous experiment come into existence?

Agricultural research at Rothamsted was started by John Bennet Lawes, who was born here in 1814 and inherited the estate eight years later when his father died. After returning home following time as a student at Oxford University he set up a laboratory in one of the best bedrooms, 'much to mother's annoyance' (Catt & Henderson 1993). Lawes went on to become a successful industrialist and businessman, selling 'patent manures' based on superphosphate and other chemicals. The patent was granted in 1842, following several years of experiments on the effects of different types of fertilisers at Rothamsted. In 1843, as his fertiliser business was taking off, Lawes employed Joseph Henry Gilbert as his full-time scientific collaborator, a partnership which was to last for 57 years. This was by all accounts a classic scientific collaboration where the abilities of the two men combined in a complementary way – with John Lawes as the public-facing, 'big-picture' side of the collaboration and Henry Gilbert as the patient, careful, rigorous experimentalist (Stevenson 1989, Catt & Henderson 1993). The extent to which these two worked as a scientific double act is shown by the Royal Society awarding the two of them its Royal Medal for important contributions to Natural Knowledge in 1867 – the only time in the award's history that it has been shared like this.

Of the various long-term experiments that they set up, probably the most famous among agriculturalists is the Broadbalk Winter Wheat Experiment – started in 1843 and still running (both this and the two Rothamsted 'wildernesses' have already been described in earlier chapters). However, to ecologists it's the Park Grass Experiment (started in 1856) that is easily the most well-known. This was also an agriculturally driven experiment when first started, as at the time horses provided much of the power on farms, and so the production of large amounts of hay to feed them was extremely important.

While the original rationale behind the Park Grass Experiment was very applied, it seems that Lawes quickly realised that the results from this, and several other of his experiments, were highly relevant to a scientific dispute between himself and the German scientist Justus von Liebig – one of the great names in nineteenth-century chemistry. One of Liebig's research areas was agricultural chemistry, and a crucial question was where plants got their nitrogen from. Nitrogen is one of the key nutrients for plant growth, along with phosphorus and potassium. At first glance it doesn't seem that access to nitrogen should be a problem for plants or other organisms, as the atmosphere is full of it. However, as has already been described in the context of nitrogen fixation by cyanobacteria (Chapter 5), most of the nitrogen in the atmosphere is in the form of dinitrogen (N_2), a molecule composed of two nitrogen atoms tied together by a very stable triple bond. This stability makes it extremely difficult for organisms to break the molecule apart and utilise the nitrogen atoms to make proteins and other chemicals. However, a small amount of nitrogen is also found in other forms in the atmosphere, such as ammonia, and this can be used by plants.

Liebig had originally thought that because of the difficulties in using dinitrogen, plants could not get enough nitrogen from the atmosphere and so needed to access the chemical from soils too, but changed his mind a few years later, arguing strongly that soil nitrogen was not needed for good crop growth. The experiments at Rothamsted, however, were showing that nitrogen from the soil is crucial for plant growth. There is an irony here, for Lawes made his money from fertilisers that mainly delivered phosphorus to plants, but because of their dispute with Liebig, Lawes and Gilbert became very interested in nitrogen in their experiments. Indeed it seems likely that we owe the continued existence of the Rothamsted long-term experiments to this dispute. The experiments quickly answered the initial questions about the effectiveness of different approaches to fertilising crops, but it seems that Lawes and Gilbert decided to keep them running as they were such a good demonstration of the fact that Liebig was wrong and, perhaps more to the point, that they were right about nitrogen and crop growth. By the time this dispute was settled in favour of Lawes and Gilbert's view, the importance of having field experiments that had been running for many years was becoming apparent, a point noted by Lawes in 1882, so despite the expense the experiments were allowed to continue (Johnston & Poulton 2018). A good accessible account of the influence of Liebig on the Rothamsted experiments can be found in the book *Demons in Eden* by Jonathan Silvertown (2005).

PARK GRASS AND PLANT DIVERSITY

What does the Park Grass Experiment show? The field itself covers some 2.8 ha and had probably been grassland for at least 100 years before the start of the experiment. To visualise the size, a hectare is approximately the playing area of an international rugby pitch, or about 50 tennis courts squeezed together. The experimental meadow is now split into 20 main plots which receive different fertiliser treatments, with plots 1–13 started in 1856 and the other plots added later in the nineteenth century. Most of these plots are divided into subplots which have received variations on the main treatment, such as different amounts of liming to study the effects of acidification (making something like 80 subplots overall). Some have had no fertiliser added, others have just received farmyard manure, while the remainder have been treated with a range of chemicals (containing various combinations of the elements nitrogen, phosphorus, potassium, sodium, magnesium and silicon).

The experiment's great importance is the long timescale. However, the fact that it was set up in the mid-nineteenth century is also its weakness, as we would now usually consider the experimental design rather poor. It lacks adequate replication (that is, multiple plots all with the same treatment to identify the range of outcomes from a particular experimental treatment), and plots were not assigned treatments randomly – which means that similar treatments may all be in the same part of the field, making it harder to know if any differences are due to the fertilisers or an oddity of that part of the meadow. The fact that we now worry about such things is also, in part, due to Rothamsted, for in the early twentieth century the research station was also central to the development of modern ideas of experimental design and statistics, ideas that we now consider invaluable when trying to look for patterns in a complex natural world. One of the founders of modern approaches to these questions was R. A. Fisher, who worked at Rothamsted for the first part of his career, getting a job there in 1919 to apply his mathematical abilities to the large amount of data that had accumulated from the Rothamsted experiments.

Although the way the Park Grass Experiment was constructed differs from the recommendations you would find in most texts on experimental design, which are heavily influenced by Fisher and his successors, this may not be as much of a problem as most people might assume. Some ecologists are starting to think that we may have overdone the emphasis on replication in ecological experiments, and that often it may be more important to include the full range of possible variation in treatments, even if this gives you fewer replicates (Kreyling *et al.* 2018). At the very least, Park Grass shows that poorly replicated experiments can still be very influential.

Our September 2005 visit couldn't have been better timed to see the effects of 149 years of continuous experimentation, as it was just before the meadow was due to be cut for hay, and so the differences between the experimental plots were clearly apparent. To a visiting naturalist plot 12 stood out as visually spectacular, dotted yellow, purple and white by a diversity of wild flowers, while other nearby plots were distinctly less colourful. For example, the next-door plot (plot 11) had few colourful flowers but had much taller grasses (Fig. 173). Looking at these plots

FIG 173. Plots 11 and 12 in the Park Grass Experiment, photographed in September 2005. The flower-rich plot 12 (to the left) has had no mineral fertiliser or manure added since the start of the experiment. Plot 11 (to the right) has had additions of nitrogen, phosphorus, potassium, sodium, magnesium and silicon.

as a conservationist, the colourful plot 12 seemed clearly preferable, but with the eye of a farmer wanting to maximise hay production plot 11 had a lot going for it. As we walked round the experiment that sunny early autumn afternoon, plot 9 also stood out, but for exactly the opposite reason from the flowery plot 12 – there were no wild flowers and only a very short extremely species-poor grassy vegetation with a few oak saplings starting to grow, presumably from acorns cached by birds the previous autumn (Fig. 174). Across the 20 plots of the experiment, plant species richness now varies from 3 to 40+ species per 200 m². Even the most species-rich plots have lost some species since the experiment was started, although the proportions of grasses, legumes and other species have stayed remarkably constant in some of the plots that have remained unlimed (Silvertown 1987).

The fact that the unfertilised flower-rich plot had relatively short vegetation while the heavily fertilised plot next door had much taller grass but fewer species is typical of the results from plots across the experiment. Plots with greater productivity, that is producing more hay, have a lower number of species growing in them. In fact this result was very quickly apparent only a few years

FIG 174. Plot 9 in the Park Grass Experiment, again photographed in September 2005. In the foreground is very species-poor vegetation growing on soil acidified by the long-term addition of ammonium sulphate as a source of nitrogen. The taller vegetation behind this short turf has had the same nutrients added but with the addition of lime to counteract this acidification.

after the start of the experiment, with very different numbers and types of species growing in the different plots, and grasses coming to dominate plots where additional nitrogen was added. This relationship between productivity and species richness is interesting, as it's often been suggested that plant communities with more species in them may tend to be more productive, as different species have somewhat different requirements and so can make better use of the resources available at a site. A useful analogy is with shops on a traditional high street – there is only enough trade for a certain number of clothes shops, but scope to increase the number of shops if you add other types, such as food shops, book shops or hardware stores. So adding more types of shops increases the overall retail productivity. Under the conditions of the Park Grass Experiment this potential positive relationship between diversity and productivity is at best rather weak.

Some plots, however – such as plot 9 – have very few species in them. Not only are parts of this plot a very species-poor short turf – in fact it's almost entirely one species, Sweet Vernal-grass *Anthoxanthum odoratum* – the soil is odd too. Kneel down and stick your fingers into a corner of the plot and you can almost peel up the sward like a carpet. Underneath the grass is a shallow peaty layer of soil unlike anything else on the meadow. If you measured the pH of this soil you would find it surprisingly acidic too, with a pH of 3.8 (the original pH of the soils here was around 5.4–5.6). The reason is that it has been treated with ammonium sulphate, a nitrogen-rich fertiliser, and on subplots where lime hasn't been added as well this has led to the very acid soil. As the soils in the area are naturally only slightly acidic, the local flora is short of species that can grow in these conditions – hence the monoculture of Sweet Vernal-grass (this illustrates the idea of environmental filtering, which I return to at the end of this chapter).

What about the diversity of organisms other than plants? Across the meadow there are some positive relationships between plant diversity and invertebrate diversity – for example, both leafhoppers and springtails are more diverse in plots with more plant species. This could be because plant diversity affects invertebrate diversity, or it could be that the experimental treatments are affecting both plants and invertebrates directly. The acidified plots, such as the non-limed parts of plot 9, are lacking in earthworms, but the experimental treatments also affect the soil microbes. Acidity appears key to microbial diversity in the Park Grass soils. Soils with low pH (that is, acidic soils) are dominated by fungi, with bacteria playing a lesser role, while in the higher pH soils bacteria become more important. This fundamental role of pH in determining what types of microbes dominate soil processes is not just restricted to the Park Grass Experiment but has been reported in many studies of soil microbial ecology. Nitrogen levels themselves seem to affect microbial diversity to a much lesser extent than they affect plant diversity in the meadow. Although there was no evidence of an effect of plant diversity on microbial diversity, increased plant diversity was associated with an increase in microbial biomass – i.e. the total mass of microbes in the soil (Rousk *et al.* 2011, Zhalnina *et al.* 2015). However, Park Grass is just one meadow in southern England – albeit an especially well-studied one. What about meadows in general?

MEADOWS – PEAKS OF PLANT DIVERSITY

In the mid-nineteenth century, when the Park Grass Experiment was started, species-rich meadows were common in many parts of Britain; however, they are now rare and often the focus of conservation management to try and maintain their diversity. Technically a meadow is grassland that is mown for hay, which means that grazing animals have to be kept off the meadow earlier in the year to allow the plants to grow. On occasions meadows cut for silage can also maintain much of the conservation interest usually associated with a traditional hay meadow. Once the meadow has been cut, animals are often grazed on it during the late summer through to early winter, and sometimes also in spring. The resulting meadows can be very rich in species, at least in plants. This profusion of flowers can attract a range of pollinating insects such as butterflies and bees, but the regular disturbance from mowing can make them difficult habitats for many animals – although for a few species, such as the Corncrake (Fig. 175), they are an important habitat (Peterken 2013). This means that in general it is plant conservationists who put a lot of effort into the preservation and management of hay meadows, although the colourful flowers and historic interest make them of concern to landscape conservationists too. Being born and brought up in Manchester, and having a fondness for the uplands, the species-rich hay meadows that come most readily to mind are those of the Pennines and Cumbria. These can have up to the low forties of plant species in a 2 × 2 m sample (usually referred to as a 'quadrat' in the technical literature), while some wetter meadows in southern Britain can have even greater plant diversity than this (Peterken 2013).

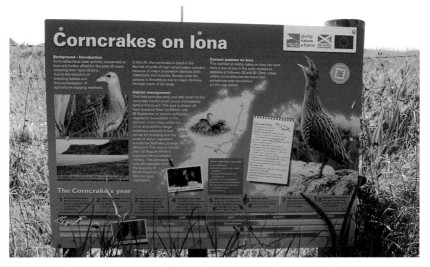

FIG 175. Meadow management (in this case for silage rather than hay) for Corncrake *Crex crex* on the Hebridean island of Iona. In the late nineteenth century this bird was found as a breeding species in every county in Britain, but it suffered a rapid decline with the increasing mechanisation of, and later decline in, hay making. It's now restricted to a few areas which still have traditionally managed meadows – especially in the Hebrides and Orkney (Holloway 1996).

The fact that you can have 40+ plant species in only 4 m^2 shows that these meadows can be species-rich at the small scale as well as at the whole-field level. This small-scale complexity (Fig. 176) was brought home to me some 30 years ago at an interview for the post of research assistant to Roger Smith at the University of Newcastle, for a job working on Pennine hay meadows. Rather to my surprise he produced a seed tray from under his desk containing a fragment of hay-meadow turf, pointed at the large number of small plant species crowded into this very limited space, and asked me how I would approach describing a community like this in mathematical terms. I don't think I gave a particularly good answer, and I wasn't offered the job. As an interview question this might seem quirky, but there was an important scientific point behind Smith's seed tray – meadow plant communities provide a spectacular and challenging example of a species-rich community for ecologists to try and explain – which is why I have used them as the central example for this chapter on aspects of ecological diversity. However, if you were to ask many people to name the most species-rich plant community, the most popular answer probably wouldn't be hay meadow. I suspect many would suggest tropical rainforest. So how do hay meadows measure up against such icons of conservation?

FIG 176. Close-up of Hannah's Meadow, a Durham Wildlife Trust reserve in Teesdale, showing the complexity of this species-rich vegetation (see also Fig. 181).

Plants are in general more easily counted than other organisms, so they are the group for which we have the best chance of answering this sort of question. A few years ago Bastow Wilson and three colleagues (2012) spent a decade noting down all relevant numbers from any research papers on plant ecology which they read, and then they used these data to try and identify the community that held the world record for plant species richness. Their answer was that there really isn't a single answer to this deceptively simple question. It all depends on the scale at which you ask it. At a large scale (100 m² and upwards) tropical rainforests do indeed 'win'. For example, 942 plant species were recorded from 1 ha of rainforest in Ecuador – to put this in context, the whole of the British Isles has only c.1,560 native plant species (Stace & Crawley 2015). However, at scales less than 100 m² all the most species-rich vegetations were semi-natural grasslands, mainly meadows. A couple of their examples came from Argentina, but most of these semi-natural grasslands were in central or eastern Europe (Figs. 177 & 178). Meadows here are even more species-rich than the British examples – for example, 43 species from only 0.1 m² at one site in Romania. All of the grasslands which featured in Wilson and colleagues' statistics were subject to periodic disturbance from human management by mowing, grazing or fire. In contrast the tropical rainforests were all apparently natural, unmanaged vegetation.

WHAT MAKES MEADOWS SPECIAL – THE INTERMEDIATE DISTURBANCE HYPOTHESIS

Why is this small-scale diversity in hay meadows surprising? Ideas about competition (Chapter 6) might suggest that a few species would do best in the conditions in any particular field and outcompete the others, so what stops this from happening? In addition, related ideas about ecological niches (Chapter 10) would seem to imply that the number of niches should limit the number of species. Are there really 40+ niches for plants in only one or two square metres of meadow? This is in fact one of the classic problems in ecological theory already discussed in Chapter 10 and often referred to as the 'paradox of the plankton' after the famous example used by Evelyn Hutchinson (1961). The issue was the large number of species of plankton that can be found in apparently uniform bodies of water, where it doesn't look likely that many different niches are available to explain this diversity. Hay meadows appear to be a terrestrial version of the same problem.

Hay meadows are interesting habitats for thinking about these questions because of their high plant diversity. What makes these meadows special?

FIG 177. Hay meadows in the Carpathian Mountains, Romania. These traditional hay meadows have high levels of plant species richness – typically with around 40 species per square metre, but this can go up to 50 species/m². Recent changes in Romania have led to the abandonment of some traditional agriculture, endangering the survival of some of these meadows (Csergő *et al.* 2013). Both photographs were taken in 2011. Top: these haycocks are on traditionally managed agricultural land associated with a monastery. Large cocks like these, supported by a central pole, are used for longer-term storage of hay. In Britain smaller more temporary cocks were often built to protect drying hay from poor weather – a practice that survived into the twentieth century, especially in the wetter north and west of the country (Peterken 2013). Bottom: flower-rich subalpine hay meadow in the Carpathians.

FIG 178. Meadows from the Dolomites. Top: close-up view of a traditional alpine meadow in the Italian Dolomites. Bottom: because of the steep terrain many meadows are still managed by fairly traditional methods – using machinery such as the small mechanical cutter seen in the photograph rather than large modern tractors.

One factor that marks them out is that they are cut for hay, so once a year the vegetation is subject to very obvious disturbance. In the context of plant ecology, 'disturbance' is often defined as 'the mechanisms which limit the plant biomass by causing its partial or total destruction' (Grime 2001). Clearly mowing, and the removal of the cut vegetation as hay or silage, meets this definition nicely. Several factors mark out disturbance as different from a stress, such as a drought. Disturbance destroys all of the species (or at least a considerable proportion of their above-ground growth), and not just those most susceptible to a particular stress. In addition, the effect is sudden – the time it takes to mow the meadow – and the effect is temporary (Wilson *et al.* 2019). One approach to trying to understand the diversity in hay meadows has focused on this regular and repeated disturbance. The approach has been called the 'intermediate disturbance hypothesis', and it is a major – if controversial – ecological idea which tries to explain diversity in relation to the frequency of disturbance events experienced by the organisms.

The basic idea is usually illustrated with a simple humpbacked graph – reminiscent of the 'limits of tolerance' graph in Chapter 2 (Fig. 16). At low levels of disturbance a small number of species do well and outcompete the other species, so diversity is low. At high levels of disturbance – think of a domestic lawn which is being cut once every week or two during the summer – conditions are tough and only those species that can put up with repeated disturbance can survive. So high levels of disturbance often lead to low diversity, which is why most naturalists find a well-manicured lawn a rather disappointing and boring habitat. A similar pattern can be seen with soil (or water) productivity. The Park Grass Experiment shows that at high productivity – that is, high fertiliser use – diversity is low, as a few species that do well under these conditions outcompete all the others. At very low productivity only a few specialist species can survive, so diversity is low too. In both cases, disturbance and productivity, we expect to see the highest diversity at intermediate values, as shown in Figure 179.

This intermediate disturbance hypothesis is an important idea which has in the past been described as 'one of the best-accepted principles in ecology' (Hoopes & Harrison 1998), and the plant ecologist Paul Keddy (2005) listed it as one of his six important 'pragmatic models' of use to plant conservationists in understanding the effects of management on plant communities. In most textbooks the idea is attributed to a 1978 paper by Joe Connell published in the leading research journal *Science* (this is the same Joe Connell whose work on competition on rocky shores is described in Chapter 6). Connell's paper contains a humpbacked graph similar to Figure 179. However, as so often in science, the textbook 'first' outing of this idea wasn't really the original description. A few

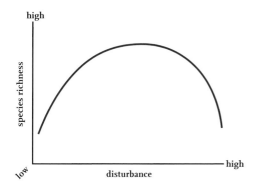

FIG 179. The classic humpbacked graph illustrating the intermediate disturbance hypothesis. Species richness peaks at an intermediate value of disturbance. The relationship between species richness and productivity is also often depicted as a similar humpbacked graph.

years earlier Henry Horn published effectively the same graph, and the earliest graphical representation of the idea appears to be in two 1973 research papers by Phil Grime which related species richness of plants in grasslands around Sheffield to both intensity of disturbance and 'stress' (such as low soil nutrient levels). However, the idea of some sort of less well specified relationship between species richness and disturbance events goes back to at least the 1940s (Wilkinson 1999b). So not only does the intermediate disturbance hypothesis apparently help to explain species richness in hay meadows, but what seems to have been the first modern statement of the idea was inspired by studies of British vegetation, namely grassland around Sheffield.

The basic idea of the intermediate disturbance hypothesis, and the similar pattern between productivity and species richness, seem intuitively sensible, and certainly appear to match what you see in grassy and herb-rich vegetation in the British countryside. Indeed, for most of my academic career (which started in the 1980s) they have tended to be treated as useful and important ideas in the textbooks. However, over the last decade or so these ideas have been challenged – with some high-profile studies suggesting they should be abandoned altogether (e.g. Adler *et al.* 2011, Fox 2013). The criticisms of these ideas have been based on perceived problems with the logic underlying the theory, and on arguments about the extent to which real plant communities show anything resembling the humpbacked graph. Why are some people sceptical?

An hour or two's search through relevant research papers, be it online or in a traditional hard-copy library, quickly illustrates some of the difficulties. For a start, there is a huge number of relevant scientific papers to look at (currently many hundreds of papers a year), many of which present results consistent with the idea, but also many that fail to find any hint of a humpbacked graph. What to make of this? In a complex world it's not surprising that the pattern isn't always seen, and, as pointed out by the early-twentieth-century astrophysicist Arthur Eddington in an oft-quoted remark, 'these experimentalists do bungle things sometimes' – so some of the studies are no doubt wrong. But how often do you have to fail to find the pattern before you start to conclude that the idea should be abandoned altogether? One sensible approach is to try to collate the results from a large number of studies and then apply various statistical techniques to these data to see how commonly they show the pattern that would be expected if the intermediate disturbance hypothesis was a reasonable description of reality. Many studies have attempted this, but it's not as straightforward as it sounds. There is much room for argument about the interpretation – for example, do these data accurately cover the full range of variation seen in nature, and have the most appropriate mathematical techniques been used in their analysis? One well-known early study that tried this approach found a convincing relationship in only 16 per cent of their data, but found stronger relationships in certain types of system, for example where the disturbance was happening in time rather than space (as in the annual cutting of a hay meadow) (Mackey & Currie 2001). Many other studies have found similar results – that is, it's hard in these big statistical analyses of large data sets to find a clear pattern supporting the intermediate disturbance hypothesis.

How do we explain this? Should we perhaps completely abandon the intermediate disturbance hypothesis, as has been suggested by Jeremy Fox (2013)? It's confusing, but there are hints of an answer. Indeed, readers with some technical background in the topic may think that my explanation so far is unnecessarily confused in describing two humpbacked graphs (one for disturbance and one for productivity) together – surely they are different and should be described in separate sections of this chapter? However, there are good reasons to put them together. As Michael Huston (2014) has pointed out, diversity and productivity are often related in nature, but most tests of these ideas have failed to address the fact that there will often be variations in both disturbance and productivity within the same study. For example, two sites that appear to have the same frequency and magnitude of disturbance may differ in soil chemistry. In addition, because competition plays a key role in the explanation of the humpbacked graph, the species being studied need to be actually competing with each other – which seems very plausible for the plants in a hay meadow but isn't always the case in studies which claim to be testing the intermediate disturbance hypothesis. In addition, Huston has also pointed out that many of the critics have misunderstood these ideas. They have criticised the ideas for failing to explain the long-term stability of diversity, when in fact they were put forward as empirical approximations that described diversity at *a particular location* in time and space. They are theories that aim to describe 'snapshots', not long epic movies, and as simple graphical ways of starting to understand a few frames of the movie they still seem rather useful. Indeed Paul Keddy (2005) has argued that such simple pragmatic models are often far more useful in guiding conservation and other management than highly sophisticated mathematical theory which can often be hard to apply in practice. I think he's probably right, but not all ecologists would agree with this assessment that there is utility in the intermediate disturbance hypothesis.

WHAT ELSE MAKES MEADOWS SPECIAL?

Whatever your view of the effects of intermediate disturbance, it's certainly not the whole answer. A range of other mechanisms contribute to the species-rich vegetation found in some British hay meadows. These not only add to our understanding of meadow diversity, but also suggest the types of ideas we may be able to apply to a much wider range of other habitat types. I will illustrate these with particular reference to the species-rich meadows of the Pennine dales and Cumbria. In a previous book in the New Naturalist series Michael Proctor described these meadows as he first knew them in the mid-twentieth century as exhibiting a 'spectacular wealth of flowers' including 'abundant Wood Crane's-bill (*Geranium sylvaticum*), Great Burnet (*Sanguisorba officinalis*) and Lady's-mantle (*Alchemilla glabra*)' (Proctor 2013). My own earliest memories of these meadows are from 20 years later, sometime in the 1970s. A friend of my father lived in the Swaledale village of Muker and we sometimes stayed in his cottage. The meadows running down from the village to the river were, in early summer, vibrant with the pinkish-purples of crane's-bills and the yellows of great seas of buttercups. No doubt these images are enhanced by the tendency for the sun to be always shining, and the flowers always blooming, in memories of childhood summers – but today these meadows can still astound, and when they are at their best in late May and June they are a perfect location for botanical tourism (Fig. 180).

These flower-rich Pennine meadows are the product of many years of traditional agriculture. Today they are mainly found towards the heads of the dales – until widespread car ownership these were very out-of-the-way places, and my father remembered that in the late 1940s and early 1950s it took all day, and multiple buses, to get from Manchester to upper Swaledale. This remoteness meant that 'modern' agriculture was slow to reach these areas. One way of attempting to understand why the meadows here are so species-rich is therefore to try and understand the past agricultural management of

FIG 180. Swaledale. Meadows at Muker in late May 2019, with the yellows of buttercup flowers (mainly Meadow Buttercup *Ranunculus acris*) and the whites of umbellifers – especially Pignut *Conopodium majus*. Swallows *Hirundo rustica* were swooping low over the meadows, Oystercatchers *Haematopus ostralegus* 'pipping' down by the river, and a Cuckoo *Cuculus canorus* was calling in the distance.

these sites (see Fig. 181 for an example from Teesdale). If the disturbance of mowing is important, then it's also important to know at what time of year the meadows were historically cut. This is a question of some practical conservation interest too, as an obvious management strategy to preserve diversity would be to continue this management into the future. Studies of farm diaries from the period 1947–1986 from various northern Pennine farms – including in upper Swaledale – show a number of interesting patterns (Smith & Jones 1991).

Two conclusions stand out from these analyses. First, the date at which hay is cut can vary tremendously from year to year because of the weather. Second, the length of the hay harvest decreased during the second half of the twentieth century. For example, at Usher Gap, which is about 1 km up the dale from Muker, the average (mean) date for the start of hay making in the 1950s – at the time Michael Proctor remembered a 'spectacular wealth of flowers' – was 1 July and the mean finish date was 18 August, with September finish dates not that uncommon. In the 1970s, when I was first getting to know the area, the start date was unchanged (mean 1 July) but the mean finish date was now 24 July. The

FIG 181. Hannah's Meadow in Teesdale, in the northern Pennines. This meadow was farmed using very traditional methods for many years by Hannah Hauxwell, who became something of a celebrity during the 1970s following TV documentaries on her life on a remote small farm without access to any modern amenities. On her retirement some of the land was taken over by the Durham Wildlife Trust, whose management aims to continue the traditional agricultural practices which have resulted in these spectacular species-rich hay meadows. See also Figure 176 for detail of the vegetation structure.

reason seems to have been changes in farming practices which allowed things to move more quickly once the weather was right. Climate change may have contributed too, but the experience of local farmers suggested that substantial improvements in technology were the main reason. In the 1950s human muscle power and horses were still in use alongside small tractors, but by the 1970s more powerful tractors and balers had arrived even in the dale heads. The timing of hay cutting helps select which plants can grow in the meadows, as ones that set seed later in the summer seldom have the chance to do so. For example, in 1988 (during the study by Smith & Jones 1991) seed production of Wood Crane's-bill peaked in early July, while the related Meadow Crane's-bill *Geranium pratense* didn't start producing much seed until late August. So it's not surprising that Wood Crane's-bill is a characteristic member of these northern Pennine hay-meadow communities while Meadow Crane's-bill is restricted to the field margins (Smith 1988).

Soil fertility matters too. This should come as no surprise; the Park Grass Experiment showed that adding nutrients tended to lower plant diversity. Traditionally the meadows are fertilised with animal manure – either from times of the year when the meadows are being grazed, or using manure from other parts of the farm. This has produced soils that are moderately fertile, and gives usable hay yields without raising fertility to levels where species richness substantially declines, as seen in some of the Park Grass plots. Roger Smith (1988) experimentally raised nutrient levels in the laboratory in small (25 × 25 cm) hay-meadow turfs from Cumbria. It may even have been one of these he surprised me with at that job interview. In this small-scale study he found that over three years the increase in nutrients was associated with a decline in species richness, albeit in what was a rather artificial situation. Later he built on this, leading a much larger field experiment looking at ways of restoring plant diversity in Yorkshire Dales meadows that had undergone agricultural improvement (Smith *et al.* 2003). This showed convincingly that reducing soil fertility had a positive effect on species richness, as predicted by the Park Grass results and many other studies. So avoiding high nutrient levels is important in maintaining these Pennine species-rich meadows, and reducing nutrient status is important in attempting to restore 'improved' meadows to a more species-rich past state.

Livestock grazing may have other important effects on fertility, by creating small-scale variation in soil chemistry within a meadow. Consider these data from a traditionally managed buttercup meadow in Ravenstonedale, Cumbria (Smith 1988). The mean (average) amount of nitrogen available for use by plants was 260.4 ppm (parts per million; 1 ppm is 1 milligram per kilogram of soil). However, just as the politician's statistic of 'average salary' obscures the fact that many people are much poorer than this figure, while others are astoundingly rich, this mean value of nitrogen hides a similarly interesting range of variation. In this field, available nitrogen actually ranged from 134 to 424 ppm, and there was a similar wide range of values for phosphorus and potassium too. Nitrogen was especially correlated with the vegetation, with different combinations of species being associated with different nitrogen levels. So this small-scale variation in soil chemistry is likely allowing a wider range of species to survive in the meadow. The variation in soil nutrients is probably created by dung and urine from the grazing stock. High-nutrient patches are those that have been recently fertilised by livestock, while it may have been many years since the lowest-nutrient patches were enhanced in this way. So even in one meadow there may be much more variation in soil chemistry than you would initially guess – suggesting that the paradox of the plankton may not be a good way of thinking about plant diversity in meadows – there can be a lot of spatial structure in the soil of a traditional hay meadow.

CONCLUDING REMARKS

Traditionally managed hay meadows provide a very interesting starting point for an analysis of the factors that lead to species-rich communities. At scales less than 100 m² (half the size of a tennis court) they include many of the most species-rich plant communities on the planet. They also illustrate a range of ideas of much more general interest and applicability. First, although very species-rich for plants they are less so for animals, many of which find the sudden change from tall vegetation to short in the middle of the summer very difficult to deal with. This illustrates a general point that a habitat that is 'good' for one type of organism may be less suitable for others.

Hay meadows are also very thought-provoking habitats when it comes to issues of conservation and human impacts on the rest of life on Earth. We are used to viewing our actions as usually damaging to the diversity of other organisms – yet these meadows have some of the highest small-scale plant diversities on Earth and are much prized by many conservationists, despite being the product of certain types of agricultural activity rather than 'natural' systems. In addition, given the many well-known environmental costs to meat production (e.g. Godfray *et al.* 2018), it's interesting to see a habitat greatly valued by many conservationists that only developed because of its utility to livestock agriculture. Many of the surviving examples of species-rich hay meadows are in the uplands, such as the Yorkshire Dales examples described in this chapter – however, there are also smaller surviving examples in the lowlands, not only at the Park Grass Experiment, but at other locations too (Fig. 182). The non-natural nature of meadows introduces topics I return to in the final chapter on human effects on the environment.

This chapter has described a number of processes involved in producing the high plant diversity found in hay meadows, and these are not unique to either hay meadows or plants but can be applied to a range of habitats and groups of organisms. The Park Grass Experiment clearly shows, with a longer run of data than any other study, the role of increased soil nutrients in reducing species richness by allowing a small number of species to do well and outcompete other plants. This is a well-established pattern, also seen, for example, in one of the longest-running ecological experiments in Africa, started in 1951 on the grassland of the South African veld. Here too increased fertiliser use reduced plant species richness, and just as at Rothamsted low pH also reduced plant diversity (Ward *et al.* 2017).

Meadows, with their annual cycle of mowing, provide particularly clear illustrations of the potential role of disturbance on species richness. While a

FIG 182. A lowland example of a fragment of surviving species-rich hay meadow: Muston Meadows National Nature Reserve in Leicestershire, which contains a population of over 10,000 Green-winged Orchids *Anacamptis morio*. This site is interesting as it was ploughed in medieval times, so the species-rich meadow has developed over the last 400 years or so.

'well-tended' domestic lawn illustrates how high levels of disturbance can reduce species richness, meadows suggest that more intermediate levels of disturbance may promote diversity by making it less likely that just a few good competitors dominate the community. Indeed there are several reasons, discussed in this chapter, why hay meadows may be systems where you would expect the intermediate disturbance hypothesis to work well as a partial explanation for the diversity. The range of soil chemistry which can be seen at a very small scale in some upland hay meadows – due at least in part to the grazing livestock – illustrates the potential importance of small-scale environmental variation in providing a range of different ecological niches, and so potentially increasing diversity. A meadow is often not uniform and may contain more niches than an initial look would suggest.

Park Grass and the northern hay meadows also illustrate another influential ecological idea, namely environmental filtering. The idea of such 'filtering' is that the environment at a given location will select against certain species, so helping to explain the makeup of any given ecological community. The very small number of flowering plants and insects that have managed to colonise the sea is another example of such filtering in action (Fig. 183). This means that not all species found in a region will be found at any particular site. The complication is that it can be very hard to experimentally tell the difference between competition effects and environmental filtering effects, and indeed these two processes often interact (Cadotte & Tucker 2017). In the context of the ideas of fundamental and realised niches, discussed in Chapter 10, this shouldn't come as a big surprise. This is a similar issue to the complications caused by the interactions between disturbance and productivity, described earlier in this chapter. However, we can certainly see something that looks like environmental filtering in the very acid plots of the Park Grass Experiment: these plots exclude the majority of local plant species, because acid soils are rare in the area and therefore few species able to grow in them are available to colonise the acidified plots. Only a small number of acid-tolerant species can pass through this filter. The northern hay meadows also appear to provide a nice filtering example, driven by cut date. Plants that set seed earlier in the summer (e.g. Wood Crane's-bill) do well in the meadows, but those that set seed after the fields have typically been mown (e.g. Meadow Crane's-bill)

 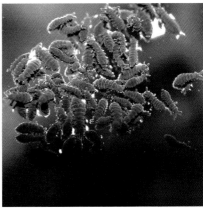

FIG 183. Two spectacular examples of environmental filtering. Although both flowering plants and true insects and their close relatives are very common on land and in fresh water, they are almost absent from marine systems. Left: a marine 'meadow' of Eelgrass *Zostera marina* exposed in a rock pool at very low tide on the Scottish island of Mull. This is one of only two species of fully marine flowering plants found in Britain (the other is the somewhat similar-looking *Z. noltei*, which has much narrower leaves). Right: the springtail *Anurida maritima* – a close relative of true insects – on the surface of a brackish rock pool at the back of a beach, also on Mull. Globally there are over a million known species of insect, but only about 1,400 are marine (Ruxton & Humphries 2008).

are unable to survive in the meadows themselves and tend to be restricted to the margins, where they may escape the harvest and so set seed.

These meadows therefore provide a colourful insight into some of the factors that are important for understanding why some habitats contain more species than others. Although they are artificial habitats created for agricultural purposes – albeit ones now rather left behind by modern agriculture – they are also ecologically informative. Indeed one can view them as accidental experiments on plant diversity, or in the case of Park Grass a deliberately constructed experiment, even though it was originally designed to address somewhat different questions.

CHAPTER 12

The View from Ringinglow Bog: Britain as a Microcosm of the Planet

I n the Peak District at 400 m on an October morning in 2017 it's unsurprising that the clouds are low, with the mist teetering on the verge of becoming real rain. The weather forecast was a bit more encouraging, and indeed things do improve as the day progresses: the cloud base lifts, with occasional bursts of sunlight glimpsed through gaps in the fast-moving clouds. From the parking at Upper Burbage Bridge a group of six of us squelch out across the tussocky surface of Ringinglow Bog, weighed down with equipment for taking cores of the underlying peat. This sort of bog vegetation is rather species-poor, and a bit of an acquired taste – one plant ecologist described the bog as being 'of almost unrivalled desolation' (Pigott 1988). However, I grew up on the fringes of Manchester with a liking for hill country, and these Peak District blanket bogs were some of the most accessible 'wild' landscapes available. So I have something of a fondness for them – despite their unnaturally low biodiversity. I am not alone in this view. Tim Lang, a well-known professor of food policy, started out as a hill farmer in Lancashire before becoming an academic and reflects that 'even though I know we must re-tree vast tracts, my spirit lifts when I see or walk on the moors' (Lang 2020). The landscapes you get to know when young are hard to view objectively in later years.

This chapter focuses on human-caused changes to the natural history of Britain. This is a subject that could form the topic of a multi-volume book in its own right, and many volumes in the New Naturalist series already cover aspects of our effects on the British environment – from early books such as *Mountains and Moorland* (Pearsall 1950) and *Man and the Land* (Stamp 1955) to more recent volumes such as *Nature in Towns and Cities* (Goode 2014), *Alien Plants* (Stace &

Crawley 2015), and *Farming and Birds* (Newton 2017). In this chapter I provide a brief overview of some key themes, especially in the context of the ecological concepts introduced in the preceding chapters. These themes include our alteration of vegetation cover over time, air pollution, the effects of introduced species, the application of ecological ideas to nature conservation and rewilding, along with the huge issue of human-driven climate change. So this chapter is necessarily more impressionistic than most of the others.

Ringinglow Bog makes a good place to start, as in the mid-twentieth century it was the subject of a well-known study of how the vegetation of the Peak District has changed over the millennia (Conway 1947). It also holds a record of more recent pollution, and indeed, as Britain was one of the starting places for the Industrial Revolution, the history of environmental changes over the last few hundred years here forms a microcosm of the changes that have now affected the whole planet.

Time-travel 70+ years back from our 2017 visit and you would see another scientist arriving at the bog, after cycling up the long hill of Ringinglow Road from Sheffield with peat coring equipment strapped to her bicycle. Her peat corer is rather antiquated by modern standards, limiting the amount of peat she can extract for laboratory analysis. The design we were using on our visit wasn't invented until the 1960s. However, the weather is similarly uncertain, and as on our visit it's a coin toss if things will improve or the clouds thicken and the rain set in. The cyclist is Verona Conway: rather reserved, even shy, and likely happier up here on the moors than in the city below. She had moved to Sheffield University a few years before as assistant lecturer in the Botany Department and started a research project on the ecology and history of Peak District peatlands (some of her earlier PhD work at Wicken Fen is described in Chapter 9). On some visits she has engineering students with her to help survey the bog surface but today she is alone – except for Miss Kolisko, who is helping with the peat coring.[6]

What Conway found is a common story in most parts of Britain. Travel back a few thousand years and the vegetation you would see around you would be very different from what is here today – even when the current vegetation looks superficially natural (Fig. 184). Humans, and especially our agriculture, rather than climate changes, are the main causes of this, leading to a country much less forested than it would have been without our actions. Conway took a core from

---

6 My reconstruction of historic fieldwork by Verona Conway collecting peat cores at the site is mainly based on her obituary by Donald Pigott (1988) and the acknowledgements section of her own 1947 paper.

FIG 184. View over Ringinglow Bog from Stanage Edge. Note all the Heather *Calluna vulgaris* and compare with the pollen data shown in Figure 185. Heather only becomes common in the time period covered by the top metre of the core. This is a relatively new vegetation at this location, one that only developed in medieval times.

the deepest part of the bog, which gave her just over 6 m of peat. Back in the laboratory she extracted pollen grains preserved in the peat samples, identified them under a microscope, and then used these data to reconstruct the history of vegetation at the site. Compare the summary of her results in Figure 185 with the general view over the bog today (Fig. 184). The heathers and grasses that dominate the modern vegetation only become really common at the top of the core (from around 1 m depth upwards), while the amount of tree pollen fluctuates, with several species becoming less common towards the top of the core and pine peaking at both the top (from modern planting, Fig. 186) and bottom (from natural pine woodland), but being much rarer in between.

Conway had no reliable way to date the peat at different depths in her core, as radiocarbon dating was only just being developed around the time she did this work, and so she had to make educated guesses at the dates. She thought the date of the rise in heather, at 100 cm depth, was probably around 1100 CE, but wasn't sure about the date of the start of peat formation at Ringinglow. A couple of decades later Sheila Hicks (1971) also studied pollen from Ringinglow. Although she also had no radiocarbon dates for this site, she did have a few from other

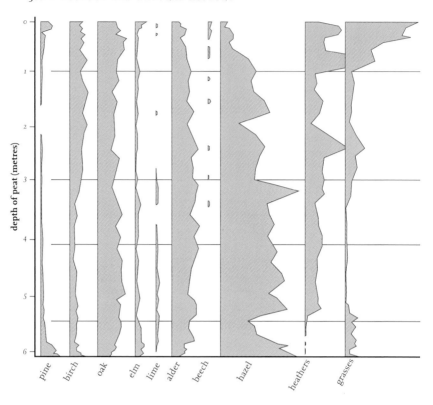

depth of peat (metres)

pine birch oak elm lime alder beech hazel heathers grasses

PREVIOUS PAGE TOP: **FIG 185.** Summary of Verona Conway's pollen data from Ringinglow Bog, showing the proportions of key types of pollen at different depths in the core (presented as percentages of total tree pollen). The horizontal lines mark out distinctive changes in the vegetation. Note that the current vegetation dominated by heathers and grasses only becomes common relatively recently. (Redrawn from Pearsall 1950)

PREVIOUS PAGE BOTTOM: **FIG 186.** Pine trees in Stanage Plantation, about 2.5 km from the core site. Note the increase in pine pollen at the very top of Verona Conway's pollen diagram (Fig. 185), associated with pine planting in the last couple of hundred years.

FIG 187. Taking a peat core from Ringinglow Bog in October 2017. The core was from close to Conway's main coring site and reached a similar depth to hers of just over 6 m – suggesting little 'growth' of peat over the last 70 years or so. Our as-yet-unpublished studies applying radiocarbon dating to this core show that a depth of around 90 cm is from the fourteenth century, while the base of our core (like Conway's, it's just over 6 m deep) is from a little over 7,000 years ago. So Conway's guess of the date of the rise in heather and grass (at around 1 m) was surprisingly good – out by only a couple of hundred years at most.

Peak District bogs, and she used these to suggest that the year 1000 was probably at approximately 60 cm depth at Ringinglow. Our recent work at the site suggests that Conway's guess was actually better than that made by Hicks (Fig. 187).

'VITIATED BY SMOKE' – THE EXAMPLE OF RECENT AIR POLLUTION

Ecology is not just a field science – often it requires the technology of the laboratory to extract the information you need from your samples. Three months on from taking the peat core from Ringinglow and it's a quiet morning in the lab, with winter sunshine slanting through a high window and falling on my back as I look down the microscope, methodically searching the slide for pollen grains that will tell me something about the past bog vegetation. The sample on the slide comes from near the top of the core – only 5 cm down – and the pollen grains are mixed in with lots of small dark particles, some spherical and others more irregular in shape. Some of these more irregular particles are microscopic bits of charcoal from bog surface fires, but much of what I am looking at is soot. There before my eyes on the microscope slide is all the dirt and grime of the Industrial Revolution, the exhalation of the dark satanic mills of Sheffield and Manchester. I am not the first to see this record of industrial pollution, for in the 1980s Jenny Jones used the upper levels of the peat at Ringinglow to study industrial pollutants (such as copper, lead and zinc) deposited on the bog surface (Jones 1987).

It's hardly a surprise to see this industrial staining; contemporary accounts of the Peak District describe extensive soot pollution. For example, Charles Moss (1913) in his account of the vegetation of the area just prior to the First World War wrote of an 'atmosphere which is frequently vitiated by smoke', which gave the plants 'a permanently dirty appearance'. The effects of all the soot can be seen on the older buildings of Sheffield and Manchester too, although this is starting to fade as old surfaces are cleaned to reveal a new less grimy exterior. Figure 188 provides a graphic illustration of the extent of former industrial pollution in Sheffield, and it was a similar story across much of the rest of the country.

The origin of this soot can be found in modern British history. In the mid-eighteenth century the British economy started on a period of rapid growth, the so-called Industrial Revolution. First it was powered mainly by water, but then increasingly by coal. The Industrial Revolution started early in Britain but quickly included countries around the world – transforming the global environment, fuelled by the combustion of fossil fuels, first coal then oil and gas (Pryor 2010). At the start of the twentieth century Charles Moss was unsure of the effects of all this smoke on local vegetation, writing that 'Whether or not the deleterious influence of smoke is a limiting factor as regards plant associations [i.e. plant communities] of the district is doubtful.' However, he went on to point out that some people had attributed declines in lichens to the smoke from burning coal (see Chapter 7 for a general introduction to lichens). The effect of air pollution on lichens became widely known to the public much later, during the 1960s, but one of the first scientific papers on the topic was published a century before by William Nylander, a reclusive Finn who lived in Paris. He described the lichens in the city's Luxembourg Gardens (Jardin du Luxembourg) – which were much healthier and more diverse than those found in more urban Paris. Nylander (1866) noted that to find a similarly good lichen flora you had to go into the countryside

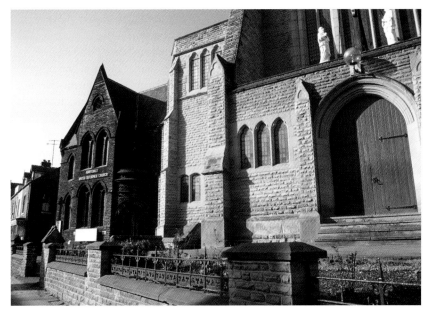

FIG 188. A reminder of the soot and grime of industrial Sheffield. These two church buildings on Abbeydale Road were photographed in 1993, when the church on the right had been recently cleaned while that to the left was still covered in soot. The second building was also cleaned not long after the photograph was taken, to reveal similarly light-coloured stone.

outside the city, and so these organisms seemed to be acting as indicators of air quality. His lichen work influenced several key British lichenologists of the mid-nineteenth century, and some of his results were translated into English (Hawksworth & Seaward 1977).

During the first half of the twentieth century it became clear that the main problem for lichens was sulphur dioxide, which mainly came from burning fossil fuels. During the 1970s the relationship between lichen diversity and sulphur dioxide became an active area of ecological research, including what we would now call citizen science projects, where schoolchildren around the country collected data on the growth forms of local lichens – although these were submitted by post rather than via the websites that would be used today (Gilbert 2000). Not all lichens suffered from the high sulphur dioxide and acidic conditions caused by air pollution. For example, *Lecanora conizaeoides* – a rather undistinguished-looking grey/green lichen – found these conditions rather to its liking. This lichen was first noticed in Britain near London in the mid-nineteenth century and quickly spread to become very common – with a similar pattern seen in continental Europe. Until the 1980s it was probably the commonest lichen in the more industrial parts of Britain, but it has now declined back to rarity again as sulphur dioxide levels have reduced in the atmosphere. As this lichen declines, other species have been expanding. Work in Germany suggests that it's a slight decline in the acidity of bark with declining sulphur dioxide pollution that is driving the loss of *L. conizaeoides*, rather than changes in parasites or competition with other lichens (Hauck *et al.* 2011).

A hint at the changes in British lichens caused by industrial air pollution comes from the Anglo-Saxon charters, dating from between the years 600 and 1080. These are legal descriptions of an area of land and use prominent features to help describe the boundaries. They include a number of mentions of 'hoar trees' – this use of the word 'hoar' survives in modern English in hoar frost, and was also used to describe the grey beards of old men. The usual interpretation is that these were trees that made notable boundary features because they were covered in trailing grey lichens such as *Usnea* and *Ramalina* (Rackham 1986). Trees bedecked with lichen like this have vanished from most of Britain and can now be found only in remoter parts of the country with no history of significant air pollution (Fig. 189). However, a more detailed glimpse of lichen diversity in

FIG 189. Branch of a hoar tree. In Anglo-Saxon times such luxuriant growths of lichen could be found in southern England, but such trees are now restricted to parts of the country with a history of low levels of air pollution. This photograph was taken in Abernethy Forest in the Cairngorms.

southern England prior to the Industrial Revolution comes from examining wood in old buildings.

Studies by Becky Yahr, Chris Ellis and Brian Coppins have investigated fragments of lichens preserved on wood in vernacular buildings in the Midlands and southern England, using wood surviving from the eighteenth century or before (Ellis *et al.* 2011, Yahr *et al.* 2011). They mainly looked at wood in the roof space of buildings, which often still had fragments of bark left on it as it wasn't worth anyone's time to tidy it up if no one would see it (Fig. 190). When they

FIG 190. Pre-industrial lichen diversity as sampled from rafters in a cottage in Downton, Wiltshire. The pole rafters (above) were made from oak and had at least 25 species of lichens growing on them when the wood was harvested. Below: close-up of bark with the leafy lichen *Parmelia sulcata* to the left (this is still a common species in lowland Britain), and various crustose lichens to the right. (Becky Yahr)

compared the species lists they had compiled from the old buildings with what is known of the current distribution of lichens in Britain, they found a substantial loss of species. There were regional variations in these data, but in southeast England the loss was around 80 per cent – which has obvious implications for a conservationist asking which species *should* be present in an area. Their analysis suggested that sulphur dioxide pollution, along with associated acid rain, was the main cause of these declines during industrialisation, but they also identified evidence for land use as a possible factor. For example, they found that although Wiltshire had suffered more sulphur dioxide pollution than Somerset it appeared to have lost fewer species of lichen. They suggested that this might have been due to the buffering effect of a larger amount of ancient woodland surviving in Wiltshire (and neighbouring Hampshire) providing extensive lichen habitat.

Over the nineteenth and much of the twentieth century the effects of sulphur dioxide in the air, and associated acid rain, have been a key factor in the ecology of lichens in Britain. This led to the almost complete absence of lichens from some of the most polluted areas, apart from a few species such as *L. conizaeoides* which could not only survive but thrive in these conditions. This provides a nice example of the difficulties in separating biotic and abiotic factors, as described in Chapter 2. Atmospheric chemistry seems to clearly meet the definition of an abiotic factor, but these chemicals were there because of the behaviour of one particular animal – an ecosystem engineer called *Homo sapiens*. However, levels of sulphur dioxide in the atmosphere have fallen substantially during recent decades and so no longer have such a profound effect on lichen distribution in Britain (except as an historical legacy, where a species was lost and hasn't yet managed to recolonise).

Other factors now affect lichens, such as the land-use changes mentioned above and climate change – for example, water availability appears to be important for lichens in urban areas in Portugal (Munzi *et al.* 2014). However, there seems to be one substantial pollution problem that now affects British lichens which has become more important as the effects of sulphur dioxide have declined. Lichens now seem to be responding to high levels of pollution by nitrogen-rich compounds in the air that make habitats more nutrient-rich. This suits some species – for example the increasingly common bright orange *Xanthoria* species – but is harmful to many others. Indeed this general increase in nutrient status of many habitats via air pollution appears to be affecting many groups of organisms, although the exact details appear complex, presumably because this pollution is interacting with many other environmental factors (Pescott *et al.* 2015). However, despite this problem, the decline in the extreme soot-laden, acidic air pollution, as described by Charles Moss in the Peak District

100 years ago and seen in the upper section of the Ringinglow Bog core, is clearly an environmental success story – albeit partial, as we still have real problems with other aspects of air pollution. No longer do you get pollution-fuelled smog in urban areas so thick that you can't see past the bonnet of your car – something that was still happening in Manchester at the start of the 1950s not long after my father had learnt to drive. Environmental successes are possible, even if they sometimes seem dwarfed by the extent of the remaining problems – one of which is the changes brought about by introduced, and potentially invasive, species.

INTRODUCED SPECIES AND NOVEL ECOSYSTEMS – A NEW 'NATURAL'?

Even with your eyes closed, this ecosystem is obviously different – the bamboos banging together under the influence of the trade wind sound like some giant garden wind chime. Open your eyes, and you see an obviously artificial pond surrounded by ferns and the bamboos that had been making all the noises (Fig. 191). This is British natural history only by the legacy of Empire – being the

FIG 191. The summit of Green Mountain on Ascension Island, photographed in 2003. Prior to human intervention the commonest large plants growing here were ferns (although the ferns visible in the photo are introduced species). A wide range of plant species was introduced during the nineteenth century (many from Kew Gardens in London) in an attempt to make the island more hospitable for the soldiers and others stationed there. This has formed an entirely novel ecosystem, a tropical forest composed of plants from around the world (Wilkinson 2004).

summit of Green Mountain on the British Overseas Territory of Ascension Island, in the tropical South Atlantic. However, it's an example of something that is common in Britain, and much of the rest of the world too, albeit usually in a less extreme form. High on Green Mountain you have to hunt hard to find any of the small number of native plant species – almost everything you see is introduced. Prior to these introductions there was no forest on the mountain at all.

Introduced plants are also important in Britain; the native British vascular plant flora is composed of around 1,560 species, but there are probably just over 2,050 introduced vascular plant species living in the wild in Britain – although the exact numbers depend on definitions used (Stace & Crawley 2015). Introduced species can sometimes dominate a habitat, as on Green Mountain: think for example of urban-fringe woodland with lots of Sycamore *Acer pseudoplatanus* and Rhododendron.

Introduced plants range from species that have become widespread while apparently causing little trouble, to others that are considered invasive and a substantial conservation and economic problem. And of course it's not only plants that can be invasive. A similar pattern can be seen with introduced species in other groups, such as the problematic rats (Fig. 192), or Harlequin Ladybirds *Harmonia axyridis*, or the Grey Squirrels discussed in Chapter 6.

FIG 192. The Common (Brown) Rat *Rattus norvegicus* arrived in Britain during the eighteenth century, largely replacing the Ship (Black) Rat *R. rattus*, which was also an introduced species that had been in Britain since Roman times. Rats have been important as a pest of stored foods, and also as a vector for some human diseases. For example, see the discussion of the role of Ship Rats in the spread of the plague in Chapter 8.

FIG 193. Pineappleweed *Matricaria discoidea*. Despite the danger sign in the background this very widely distributed introduced plant doesn't seem to cause any significant problems. Pineappleweed does well in sites with quite a bit of disturbance from trampling, such as here at the edge of a car park in Cley, Norfolk.

A familiar plant example of a widely distributed introduced species in Britain is Pineappleweed (Fig. 193). This species has been so successful in spreading round the world with our help that the limits of its original distribution are now uncertain. It was first cultivated in Britain in the eighteenth century, but seems to have escaped into the wild from Kew Gardens in London in 1871 and is now found throughout the country, with the exception of some high mountains (Preston *et al.* 2002). Edward Salisbury (1964b), whose active career as a botanist spanned the first three-quarters of the twentieth century, pointed out that its great increase in distribution in Britain was over about 25 years from 1900 onwards. He noted that the average fruit production from a single plant is around 7,000 fruits – with a very high (93 per cent) germination success. So there is considerable scope for dispersal. He attributed the expansion in the plant's range in the early twentieth century to, at least in part, the rise in motor vehicle use. Pointing out that 'it must not be forgotten that these early years of motoring were an era of dusty travel and motoring goggles, prior to the advent of the tarmac surface, and hence also the era of prolific mud'. The suggestion was that the fruits were spread widely on the tyres of early cars, and the fact that this is a plant that grows well on trampled and disturbed ground makes this seem a very plausible mechanism. In the context of this chapter the interesting point is that despite its occurrence across most of the country I am unaware of any suggestions that this species has become a problem plant.

Not all introduced species are so benign. For example, Parrot's-feather (Fig. 194), an aquatic plant once much planted in garden ponds, is clearly invasive; indeed, the problems it potentially causes are such that its sale in Britain has now been banned. It was first found growing in the wild in 1960 in southern England and – as a native of warmer parts of South America – may be favoured by climate change. The term 'invasive' is used for such species which 'can attain a substantial portion of the biomass' in invaded semi-natural plant communities (Stace & Crawley 2015). Another well-known example in Britain is Rhododendron (Fig. 151). Parrot's-feather and Rhododendron are the exceptions, as it's only a minority of naturalised non-native plant species that have become invasive in Britain: Stace and Crawley (2015) estimate it's only around 5 per cent that acquire invasive status. However, it has proved anything but straightforward to predict in advance which will be the problematic species. One of the many complications may be mycorrhizal fungi, for some introduced plant species seem to be able to form relationships with fungi which are already present while others may need their fungi to arrive with them (Vlk *et al.* 2020).

Introductions are not restricted to plants and animals, for many introduced fungi are thriving in Britain too. Some, including plant pathogens such as the

FIG 194. Parrot's-feather *Myriophyllum aquaticum*, a plant of central South America, was first recorded growing in the wild in Britain in 1960 (Preston *et al.* 2002). As can be seen in the photograph it grows in dense masses emerging from the water, and can shade out other aquatic species.

fungus that causes ash dieback (see Chapter 8), are clearly problematic, while others appear similar to Pineappleweed in adding to British diversity without apparently causing many problems. An intriguing example is *Agrocybe rivulosa* (Fig. 195). This species was first described in 2003, based on specimens found growing on woodchip in the Netherlands (Nauta 2003). It was first found in Britain in 2004 in Staffordshire, and within only five years it was being widely recorded from many parts of the country. To what extent this was due to a very rapid spread, or whether it was partly due to naturalists starting to look for it, is uncertain. The original range of this species is unknown, but guessed to be tropical or subtropical (woodchip piles can be warmed by microbial decay processes). The likelihood is that it has been moved around the planet by human action. However, its spores are of a size that potentially allow wind dispersal between continents, so a more natural spread cannot be completely ruled out (Wilkinson *et al.* 2012b). Now a characteristic mushroom of woodchip piles in

FIG 195. *Agrocybe rivulosa* growing on woodchip at the Centre for Alternative Technology near Machynlleth in southern Snowdonia, November 2015. This fungus doesn't appear to have acquired an English name, but the English translation of the Dutch common name is veined clay hat. For a description of the discovery of the first British occurrence of this species see Lovett (2006).

Britain, *A. rivulosa* is not obviously a problem, and in addition illustrates just how little we know of global fungal diversity when a species can be first described by science as an introduction in Europe, rather than from its presumed original home in the tropics.

Systems such as the summit of Green Mountain are increasingly common around the world and have started to be referred to under the term 'novel ecosystems' (Hobbs *et al.* 2013). These are self-sustaining ecosystems which have no historical precedent – for example because they are composed of non-native species or formed under new climatic conditions. These differ from slightly modified historical ecosystems, which with some management can potentially be returned to something like their original state. Consider the summit of Green Mountain – as almost all plant species growing there are non-native it's not practical to return it to the conditions before those plants were introduced. A British woodland dominated by Sycamore and Rhododendron is somewhat similar – remove these species and effectively you have no woodland left. Pragmatically, with so many ecosystems profoundly changed by human actions, novel ecosystems are here to stay. Many systems are not yet changed to this extent, however, and it is therefore important to ask questions about the role of introduced species within these hybrid systems – that is, ecosystems part way between historical systems and novel systems (indeed, such hybrid systems now make up much of Britain, including most nature reserves). In principle it may be possible to remove many of the introduced species from these systems, but is it always worth the time and expense of such management?

At a local scale some invasive plants clearly outcompete other species and can come to dominate, as with the Parrot's-feather in Figure 194. But what if we ask the question at larger scales? Do invasive plant species do much damage at a national scale? One attempt to answer this question for Britain used data from the British Countryside Survey for 479 sites, looking at changes between 1990 and 2007 (Thomas & Palmer 2015). This work found that at locations where non-native species were increasing the native plant species tended to be increasing too. Chris Thomas, who was the lead author in this study, has become a prominent sceptic of the idea that introduced species are necessarily bad for conservation (Thomas 2017). Chris likes to challenge his colleagues to name a species that has gone extinct in Britain owing to the effects of invasive species. When we had this conversation a few years ago the best we could come up with was the introduction of myxoma virus (arguably not actually a living organism) reducing the numbers of Rabbits *Oryctolagus cuniculus* (themselves introduced), and so leading to habitat changes that contributed to the extinction of the Large Blue Butterfly *Phengaris arion* (studied in detail by Jeremy Thomas, Chris's brother). However, it's possible to think of examples where things could have turned out differently. An example is Lundy Cabbage *Coincya wrightii*, a species whose entire global range is restricted to the island of Lundy in the Bristol Channel; it also has at least one endemic beetle species associated with it. This plant is only found on the southeast cliffs of the island and these are threatened by the spread of Rhododendron, first introduced to the island in the early nineteenth century (Compton *et al.* 2002). So without management to control Rhododendron the Lundy Cabbage could potentially have been an example of an endemic British species that became extinct because of an invasive non-native.

ECOLOGY AND NATURE CONSERVATION

Cley Marshes, North Norfolk, early June 2018. The view through my binoculars would have astonished my teenage self, 40 years ago. In a single field of view are multiple Avocets *Recurvirostra avosetta* and a Spoonbill *Platalea leucorodia*; Little Egrets *Egretta garzetta* are common too, and even in the winter you can watch Avocets (the numbers of Avocets in Britain in winter having expanded greatly during my lifetime). These Norfolk marshes, with their spectacular waterbirds, have been a nature reserve since 1926, and have already been briefly used as an example in this book in the chapter on competition (Chapter 6). Clearly many of the ideas in this book can be applied to management for nature conservation, such as the concepts of succession, intermediate disturbance and metapopulations. Some conservation-related examples are included in earlier chapters, but here I briefly discuss a few wider issues before considering the all-embracing issue of climate change.

All these Norfolk birds had become extinct as breeding species in Britain, but have now returned to breed. Spoonbills bred at several locations in southern Britain until the seventeenth century, while historical records seem to suggest that Little Egrets were widely eaten in medieval times – although if this is correct it's odd that there are only very few archaeological bones of this species from Britain (Yalden & Albarella 2009). Management of wetland sites for conservation during the twentieth century, with associated protection from hunting, has led to their return. Reedbeds such as those at Cley (Fig. 83) and other sites around the country (Fig. 196) nicely illustrate the importance of ecological succession, as they are midway along a successional sequence (Fig. 142), and to maintain them on a nature reserve usually requires periodic cutting to prevent succession to fen and other vegetation types. This maintains a mosaic of reed and more open-water habitats.

FIG 196. Reedbeds and Great White Egret *Ardea alba* at the RSPB's Leighton Moss reserve in Lancashire. Continental European populations of this egret are expanding, and it's now seen much more frequently in Britain.

At the small scale of a garden managed for wildlife, up to the larger scale of a nature reserve, the diversity of species can often be enhanced by management to increase the number of habitats and niches available. However, by doing this you necessarily decrease the extent of any one habitat type, which may make the site less suited to some habitat specialists that need larger areas of a particular habitat to survive. This can clearly be seen at the very small scale of wildlife gardening. For example, when I recently put a reasonable-sized pond in my garden, this had the effect of decreasing the amount of space for meadow-type vegetation (Fig. 197). Clearly this sort of manipulative conservation management focused at a site level (sometimes disparagingly described as 'conservation gardening') can lead to important improvements. The down side is that it's obvious that most of the country can't be managed primarily as a nature reserve, so larger landscape-level management is needed.

One high-profile approach to conservation at the large scale is the idea of rewilding – briefly described in Chapter 9 in the context of succession. A well-known British example of this approach at a farm scale is Knepp in West Sussex. This 1,400 ha mixed farm was converted to a more wild state with free-ranging Longhorn cattle and other grazers – developing a vegetation of 'thickets, scattered

FIG 197. My garden in Nottinghamshire. Sweep-netting the long grass during summer produces large numbers of stilt bugs, shield bugs and spiders. The mix of habitats – pond, short grass, long grass, trees and shrubs – increases the diversity of species that can live in the garden but necessarily reduces that amount of any given habitat. For example, digging the pond reduced the amount of space available for meadow vegetation.

trees and long grass' and a considerable natural history interest (Marren 2016). West Sussex is a relatively crowded county with high land prices, but there is scope for much larger rewilding schemes in the British uplands – such as the relatively species-poor moors of the Peak District described at the start of this chapter. Rewilding itself is a somewhat nebulous term meaning different things to different people, from reintroducing long-lost charismatic species, to allowing more natural processes (such as succession) to play a larger role than in conventional management (i.e. leaving it to nature). All these approaches are somewhat problematic once you start to think about them in more detail. For example, when I started giving university lectures in the early 1990s I would often talk about using approaches such as pollen analysis to try and establish historical conditions which could be used as a guide for conservation management – indeed, I often used the pollen diagram from Ringinglow Bog (Fig. 185) as one of my examples. However, in the context of a changing climate the usefulness of historical conditions as conservation targets looks increasingly problematic (Harris *et al.* 2006). At the same time, if natural succession is left to create future ecosystems in Britain these will likely have a large component of introduced species, something that still makes many conservationists very uneasy.

At a national scale, any substantial rewilding cannot be discussed in isolation, as a purely conservation/environmental question. In a country that currently only produces around half of the food it requires, agriculture needs to be discussed alongside conservation. Much of the environmental damage done to feed the British population is currently outsourced to other countries, a situation that doesn't look very sustainable and one we need to improve on (Lang 2020). At a very small scale this problem is illustrated by the garden shown in Figure 197 – because most of the garden is managed for its natural history interest there is very little space available for growing human food. Despite these problems, as Tim Caro (2007) has pointed out, there are obvious attractions in the idea of rewilding:

> *Conservation biology has developed into a science of documenting population declines, species losses and habitat destruction … re-wilding is a proactive idea that could galvanize the conservation community out of its helplessness and, for that alone, deserves merit.*

A central objective of conservation is to prevent species extinction – be that locally in a region, or total global extinction. The list of birds in Norfolk that have returned as breeding species gives a rather upbeat view of British conservation; however, there has also been a long list of extinctions in Britain over the last couple of centuries. Clive Hambler and colleagues (2011) estimated that between 1 and 5 per cent of species (of plants, animals and everything else) have been lost from Britain per century over this time period – with the rate of extinctions probably greater in the twentieth century than the nineteenth. Groups that saw higher rates of extinction include ones associated with dead wood and aquatic habitats, along with species that were close to the edge of their climatic range. One group that has attracted particular interest are bees and other pollinating insects, because of the potential effects of any decline on ecosystems and agriculture. Britain has good historical records of many insect species because of the long-term interest in natural history in the country. An analysis of these records for bees and flower-visiting wasps showed multiple extinctions from Britain from the mid-nineteenth century until the end of the twentieth century, and many of these seemed to be associated with changes in agriculture from the 1920s onwards (Ollerton *et al.* 2014).

Do such extinctions matter to ecosystem functioning? A common experimental approach to this question has been to randomly remove species from plant communities (plants make easier experimental subjects than animals) to see what effect this has on various measures of ecosystem function. However, in reality species are not lost randomly, and some are more likely to be lost than others – for example in Britain those from dead wood and aquatic habitats, as already mentioned. In addition, some species are potentially more important than others – such as ecosystem engineers and keystone species (described in Chapter 10). Loss of key pollinators may be an example of the effects of losing keystone species. Another reason why a particular species may turn out to be especially important is if it is particularly dominant in the ecosystem – therefore having a particularly large influence on ecosystem properties such as the production of biomass. This idea, called the mass ratio hypothesis, was developed for plant communities by Phil Grime (1998), and recent studies of North American prairie grasslands suggest that this mass ratio effect is indeed important. How many species are lost may be less important than the identities of those lost (Smith *et al.* 2020). Because they are by definition common, it may seem that the loss of such species is unlikely. But this may not always be the case – think of the loss of common tree species to introduced plant pathogens.

CLIMATE CHANGE AND THE EARTH SYSTEM

Return to Ringinglow Bog and think about our peat core, this time in the context of climate change rather than reconstructing the history of the environment. Our core came from a location on the bog close to where Conway took one of her main cores – after extensive coring and surveying she had identified this as the deepest part of the peat bog. Both her core and ours sampled peat just over 6 m deep.

Peat is close to 100 per cent organic matter. The classic way to measure the amount of organic matter in a peat or soil is to burn it at very high temperatures in a furnace (referred to as loss on ignition). The organic matter turns to carbon dioxide and only the inorganic material is left behind in the crucible. The difference in weight between the dried sample and the sample after combustion gives you an estimate of the amount that is made up of organic matter. In a normal mineral soil, often plenty of the sample is still to be seen in the crucible after ignition in the furnace, but for a true peat only a small trace of inorganic dust remains – most of the sample has turned to carbon dioxide. So a deep bog such as Ringinglow contains large amounts of the greenhouse gas carbon dioxide locked up in its peats. Indeed, over the last few decades the importance of peatlands as carbon stores has become a major factor in arguing for their conservation, to go alongside the specialist plants and animals which have peatlands as their habitats. What is clear is that if all the carbon in the peat from bogs of upland and northern Britain were to return to the atmosphere it would add substantial amounts of carbon dioxide, with serious implications for climate change.

A complication that needs mentioning here is that peatlands are also a source of another greenhouse gas – methane – produced by microbes in the waterlogged conditions. So they lock up one greenhouse gas but release another. A further complicating factor is that the ability of peat to lock up carbon can be compromised by interactions with pollutants: for example, the acid rain of former industry killed off most of the peat-forming bog mosses *Sphagnum* spp. downwind of major urban areas, reducing the rate at which new peat forms and so the bogs' ability to lock up additional carbon dioxide. The southern Pennines – including Ringinglow – suffered some of the most extreme acid rain in Europe, equivalent to one litre of concentrated sulphuric acid on every square metre between 1880 and 1991 (Skeffington *et al.* 1998). Few bog mosses survived this chemical assault.

Human-driven climate change not only undermines the logic of using past conditions as a model for conservation objectives, but is also a major factor affecting a very wide range of issues in applied ecology. In fact, the difficulty is of thinking of questions *not* affected by the prospect of ongoing global heating. The potential importance of this issue is underlined by my preference for the term global heating rather than the more widely used global warming. Heating better describes the process by which we are altering the energy balance of the planet – so leading to the well-known warming effect (Box 3). The term 'global heating' has been picked up and used by some commentators (such as the *Guardian* newspaper in the UK), not so much because it more accurately describes planetary heat balance, but to better stress the extent of the problem, as warming sounds a rather mild process. As a term it's been around for some time but was brought to prominence by James Lovelock in his 2006 book *The Revenge of Gaia*.

If we want to make informed speculations about the likely future effect of climate change on British natural history then we have to make use of the output of climate models to identify likely changes in the British climate. However, we don't need to rely on models to identify the fact that there has been warming over the last 50 years or so – the record of actual observations shows this. In Britain the period 2008–2017 was on average 0.3 °C warmer than 1981–2010, and 0.8 °C warmer than 1961–1990. Modelling results suggest that during the current century Britain is likely to have warmer and wetter winters, with hotter drier summers (Lowe *et al.* 2018). A hint of these future summers comes from the fact that the UK temperature record was broken in July 2019 in Cambridge with a

---

BOX 3. Global heating and energy balance

(Based mainly on McGuffie & Henderson-Sellers 2005)
There are many ways of thinking about the Earth's climate in a mathematical way, and of constructing models to make predictions about the future. The simplest approach is an energy balance model. This is effectively an accounting exercise – quantifying all the energy entering the atmosphere from the sun and all the energy being lost to space from the Earth. Remember that heat is a form of energy. If this system is in perfect balance then the planetary temperature will stay constant, if more energy is arriving than being lost then the system warms up, and if more is being lost than arriving then the system cools down. Hence this approach is called an energy *balance* model. Although these models are the simplest approach to modelling the global climate they were first used only in 1969; early versions of more complex models that attempt to include atmospheric circulation had actually been used earlier, in the 1950s – and given the very limited computer storage available at the time (5 kilobytes in the first such study) that was an extraordinary achievement!

Over 70 per cent of the incoming energy that drives the climate system is absorbed by the Earth's surface (so warming the ground or water). This means that the colour of the planetary surface is important, as lighter colours reflect more energy back out to space. Ecology is important here; for example, forests tend to be dark and absorb more energy, and chemicals from marine plankton can help cloud formation (the effect of clouds is complex – depending on type, they can either reflect energy or help trap it within the atmosphere). The loss of energy to space is controlled both by the temperature of the Earth and the transparency of the atmosphere to outgoing heat energy. Greenhouse gases (such as water vapour, carbon dioxide and methane) reduce the transparency of the atmosphere to this outgoing energy and so heat the climate system. Hence global heating.

---

temperature of 38.7 °C, and such extremes are expected to become more frequent with global heating (Christidis *et al.* 2020).

One way of attempting to get an idea of what this might mean for British natural history is to look back in time to previous conditions that were warmer than today's. The recent geological past has been characterised by repeated cold (glacial) and warmer (interglacial) periods. The last interglacial is known to have been warmer than our current one, and so has attracted interest as a possible model for what things might look like in the future. However good a model it may be for future climates, it's unlikely we are about to revert to the charismatic mammal fauna found in Britain at the time (at the peak of warmth around

125,000 years ago), which included Hippo *Hippopotamus amphibius*, Lion *Panthera leo* and the extinct Straight-tusked Elephant *Palaeoloxodon antiquus*. In fact it's not these mammals but the plants, and especially the beetles, that are most informative about the climate at the time. One of the best-known sites for fossils from this interglacial is Trafalgar Square in London. Estimates based on the fossils found during building works suggest a maximum summer temperature 3–5 °C warmer than the last few decades (to give context, a rise of 2–3 °C by 2100 is thought possible without strong action to control global heating), but crucially the winter temperature at the time looks no warmer, and perhaps slightly cooler, than today. So the climate was more seasonal than we are used to, and the warmer winters predicted for Britain over the next few decades were not present in the last interglacial. The Trafalgar Square site appears to be from the very warmest part of this interglacial, with other known sites in Britain from that interglacial looking a little cooler. The plants suggest this too: Water Chestnut (Fig. 198), which seems to prefer summer temperatures of at least 20 °C, is only found in the London samples (Candy *et al.* 2016). Although not an ideal match for the future (especially in winter temperatures, and also in the absence of humans), the last

FIG 198. Water Chestnut *Trapa natans* growing in Romania. Add a few Hippos and you would have a view of the area around Trafalgar Square during the last interglacial.

interglacial does show a warmer Britain supporting a rather different collection of animals and plants than today.

If a time of Hippos and Water Chestnut in the wetlands of London 125,000 years ago seems a bit remote as a guide to future Britain, then we can turn to recent warmer years instead. Early examples of these were the years 1988–1990, which had unusually mild winters and hot summers (which were also very dry in southeast England). A Department of the Environment commissioned report at the time summarised some of the effects (Cannell & Pitcairn 1993). For example, many lawns and grassland went brown, but some species such as Ribwort Plantain *Plantago lanceolata* with deeper roots stayed green, many trees wilted and even lost leaves earlier than normal, while plants of moist habitats such as ferns suffered shoot death. The mild winters meant that the insect traps at Rothamsted were catching aphids much earlier in the year than normal, and lizards and snakes were active in both winters. This warm weather of 1988–1990, unusual enough at

FIG 199. The cliffs of Clogwyn Du'r Arddu on Snowdon, in winter and summer. Like several other Snowdonian cliffs (e.g. Figs. 13 and 65), this is a classic site for rare arctic–alpine plants. Climate change makes it likely that snowy conditions will become much rarer as winters become milder, and hotter drier summers may dry out even north-facing cliffs such as these, reducing their suitability for many montane plants.

the time for the government to commission a report on the effects, is becoming increasingly common. One example of a group under threat, used as an example on several occasions in this book because of its ecological interest, are mountain plants (Fig. 199). Trevor Beebee (2018) has recently pointed out that our mountains are 'places where concerns about negative impacts of climate change are acute', as key species found here are adapted to cool conditions. British mountains are relatively low, so there is limited scope for these species to retreat to higher altitudes as the climate warms, making them a particular conservation problem.

CONCLUDING REMARKS

The view of Britain you get from Ringinglow Bog is one of a country greatly changed by human activity, and in this respect it is a microcosm of the whole planet. Pollen grains preserved in the peat tell of large changes to the vegetation over the last few thousand years, a story repeated in the pollen record from bogs and lake sediments across the country. The top of the core, black with industrial soot, marks the major changes due to the large-scale reliance on fossil fuels associated with the Industrial Revolution. Indeed, the core contains a much longer record of industrial pollution – the work I am involved with is looking at the history of lead pollution in the Peak District, and related work from Alpine ice cores suggests that pollutants from Peak District lead production were spreading as far as continental Europe in the Middle Ages (Loveluck et al. 2020). This story of pollution from one country having effects at a continental scale even in the twelfth and thirteenth centuries is a reminder that the environmental effects of humans are not just local to where their activities are taking place, and not just restricted to the last few hundred years. The carbon locked up in the peat beneath my feet, and all the soot in the upper layers of the bog, are reminders that we are now affecting the climate on a global scale through the burning of fossil fuels and landscape changes.

This is a book on basic ecological ideas, rather than one on the many and varied ways humans have altered the planet. It is therefore interesting to step back and ask how the changes we are making to the Earth fit into the broad theoretical perspective. In the opening chapter I described the importance of free energy in ecology – all living organisms need to draw energy, and other resources, from their environment, and as their use can never be 100 per cent efficient waste products are released back into the environment. So what humans are doing is not qualitatively different from what any other organism does, it's just that we are now literally doing it on an industrial scale.

As pointed out in Chapter 1, this way of thinking about the basic concepts of ecology draws on the science of thermodynamics. In his 2016 book *Thermodynamic Foundations of the Earth System* Axel Kleidon pointed out that:

> *much of the impacts of human activity on the Earth system are caused by food production and consumption of fossil fuels. Both aspects involve the appropriation of chemical free energy from the Earth system in terms of carbohydrates from biomass or geologic deposits.*

These problems are global, and therefore require global solutions. I have already pointed out in this chapter that currently Britain outsources much of the environmental damage associated with production of the food we eat. A solution where we expand conservation within this country at the expense of the environment in other countries is no solution at all if we take a global view. The moral philosopher Mary Midgley was fond of using a metaphor of a ship to represent the Earth when thinking about global environmental issues and the importance of acting at a planetary scale, writing that 'Disasters do not respect national boundaries. Ships that sink tend to sink at both ends' (Midgley 2001). In the ecosystems chapter of this book (Chapter 5) an Earth systems approach to viewing the planetary ecosystem is described. Although it is in the nature of books in the New Naturalist series that they focus on the natural history of Britain, ultimately solutions to most of the larger problems in applied ecology require an international approach, and one that views us as very much part of the global ecosystem.

To further make this point I end with an image. The University of Lund in Sweden has a strong reputation in ecological research, and in the university square, on the site of its old eighteenth-century botanic garden, is a collection of sculptures. One of these is 'Man escaping from the rock' (alternatively, 'Man escaping from the darkness of ignorance') by Axel Ebbe, presented by the City of Lund to mark the university's 250th anniversary in 1918 (Fig. 200). It's one of my favourite visual representations of how we should view the place of humans on the Earth – not as something apart from the rest of the planet but as an animal very much embedded in the totality of life. This also illustrates a key difference between the arts and sciences. If one of my explanations in this book is open to many different readings, then I have simply failed to be clear. However, ambiguity and multiple interpretations are often a strength in art. Axel Ebbe seems to have intended this sculpture to show a person escaping the natural world into a human world of enlightenment, while I see the figure as still very much embedded in the rock of nature. In this latter reading the sculpture is a good metaphor for our relationship to the natural world, and possibly the sculptor's original intention makes a good example of where we have gone wrong. We are clearly part of the many ecological processes on this planet – as the examples used in this chapter have illustrated. To paraphrase the distinguished biologist Peter Medawar (1982), we need to use the concepts of ecological science to underpin a new technology to alleviate environmental problems, in the same way that the technology of medicine is founded upon the science of physiology.

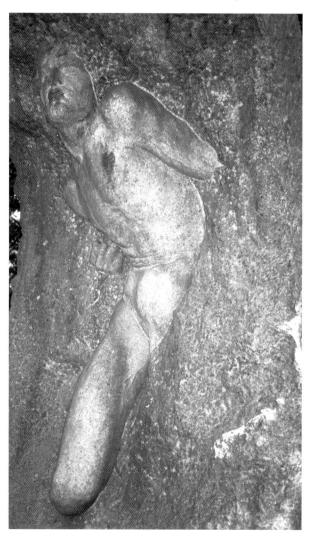

FIG 200. 'Man escaping from the rock' by Axel Ebbe at Lund University in Sweden. The sculptor seems to have intended this sculpture to show a person escaping the natural world into a human world of enlightenment, whereas, as described in the text, I see the figure as still very much embedded in the rock of nature.

References

Adamson, J. (2009). Moor House memories: a social history of a nature reserve. UK Environmental Change Network. www. ecn.ac.uk/publications/moor-house-memories (accessed October 2020).

Adler, P. B., Seabloom, E. W., Borer, E. T. et al. (2011). Productivity is a poor predictor of plant species richness. *Science* 333, 1750–1753.

Alexander, J. M., Diez, J. M. & Levine, J. M. (2015). Novel competitors shape species' responses to climate change. *Nature* 525, 515–518.

Anderson, B. (2010). Did *Drosera* evolve long scapes to stop their pollinators from being eaten? *Annals of Botany* 106, 653–657.

Anderson, R. M. & May, R. M. (1991). *Infectious Diseases of Humans: Dynamics and Control.* Oxford University Press, Oxford.

Anderson, T. R. (2013). *The Life of David Lack.* Oxford University Press, Oxford.

Andrewartha, H. G. & Birch, L. C. (1954). *The Distribution and Abundance of Animals.* The University of Chicago Press, Chicago.

Archer-Thomson, J. & Cremona, J. (2019). *Rocky Shores.* Bloomsbury, London.

Asher, J., Warren, M., Fox, R. et al. (2001). *The Millennium Atlas of Butterflies in Britain and Ireland.* Oxford University Press, Oxford.

Atkins, P. (2007). *Four Laws that Drive the Universe.* Oxford University Press, Oxford.

Avanzi, C., Del-Pozo, J., Benjak, A. et al. (2016). Red squirrels in the British Isles are infected with leprosy bacilli. *Science* 354, 744–747.

Aybes, C. & Yalden, D. W. (1995). Place-name evidence for the former distribution and status of Wolves and Beavers in Britain. *Mammal Review* 25, 201–227.

Babikova, Z., Johnson, D., Bruce, T., Pickett, J. & Gilbert, L. (2013). Underground allies: how and why do mycelial networks help plants defend themselves? *Bioessays* 36, 21–26.

Baker, J., Humphries, S., Ferguson-Gow, H., Meade, A. & Venditti, C. (2020). Rapid decreases in relative testes mass among monogamous birds but not other vertebrates. *Ecology Letters* 23, 283–292.

Bardgett, R. D., Leemans, D. K., Cook, R. & Hobbs, P. J. (1997). Seasonality of the soil biota of grazed and ungrazed hill grassland. *Soil Biology and Biochemistry* 29, 1285–1294.

Bar-On, Y. M., Phillips, R. & Milo, R. (2018). The biomass distribution on Earth. *Proceedings of the National Academy of Sciences of the USA* 115, 6506–6511.

Bates, S. T., Berg-Lyons, D., Lauber, C. L. et al. (2012). A preliminary survey of lichen associated eukaryotes using pyrosequencing. *The Lichenologist* 44, 137–146.

Beebee, T. (2018). *Climate Change and British Wildlife.* Bloomsbury, London.

Bell, A. D. (1991). *Plant Form: an Illustrated Guide to Flowering Plant Morphology.* Oxford University Press, Oxford.

Berry, R. J. (1977). *Inheritance and Natural History.* New Naturalist 61. Collins, London.

Birkhead, T. (2000). *Promiscuity.* Faber & Faber, London.

Birkhead, T. R. & Furness, R. W. (1985). Regulation of seabird populations. In R. M. Sibly & R. H. Smith (eds.), *Behavioural Ecology.* Blackwell, Oxford, pp. 145–167.

Birks, H. J. B. (2019). Contributions of Quaternary botany to modern ecology and biogeography. *Plant Ecology and Diversity* 12, 189–385.

Bishop, I. J., Bennion, H., Sayer, C. D., Patmore, I. R. & Yang, H. (2019). Filling the 'data gap': Using paleoecology to investigate the decline of *Najas flexilis* (a rare aquatic plant). *Geo: Geography and Environment* 2019, e00081.

Bowler, P. J. (1992). *The Fontana History of the Environmental Sciences.* Fontana Press, London.

Bradshaw, A. D. (1993). Introduction: understanding the fundamentals of succession. In J. Miles & D. W. H. Walton (eds.), *Primary Succession on Land.* Blackwell, Oxford. pp. 1–3.

Brooks, D. R., Bater, J. E., Clark, S. J. et al. (2012). Large carabid beetle declines in a United Kingdom monitoring network increases evidence for a widespread loss in insect biodiversity. *Journal of Applied Ecology* 49, 1009–1019.

Brown, J. (2002). *Charles Darwin: the Power of Place.* Jonathan Cape, London.

Buesching, C. D., Clarke, J. R., Ellwood, S. A. et al. (2010). The mammals of Wytham Woods. In P. S. Savill, C. M. Perrins, K. J. Kirby & N. Fisher (eds.), *Wytham Woods, Oxford's Ecological Laboratory.* Oxford University Press, Oxford. pp. 173–196.

Burke, T. & Bruford, M. W. (1987). DNA fingerprinting in birds. *Nature* 327, 149–152.

Burki, F., Roger, A. J., Brown, M. W. & Simpson, A. G. B. (2020). The new tree of Eukaryotes. *Trends in Ecology and Evolution* 35, 43–55.

Busher, P. E. (2016). Family Castoridae (beavers). In D. E. Wilson, T. E. Lacher Jr & R. A. Mittermeier (eds.), *Handbook of Mammals of the World.* Lynx Edicions, Barcelona, pp. 150–168.

Butt, N., Bebber, D. P., Riutta, T. et al. (2014). Relationships between tree growth and weather extremes: spatial and interspecific comparisons in a temperate broadleaf forest. *Forest Ecology and Management* 334, 209–216.

Cadotte, M. W. & Tucker, C. M. (2017). Should environmental filtering be abandoned? *Trends in Ecology and Evolution* 32, 429–437.

Calladine, J., Jarrett, D. & Wilson, M. (2019). Breeding bird assemblages supported by developing upland scrub woodland are influenced by microclimate and habitat structure. *Bird Study* 66, 178–186.

Campbell, B. (1979). *Birdwatcher at Large.* J. M. Dent & Sons, London.

Candy, I., White, T. S. & Ellias, S. (2016). How warm was Britain during the last interglacial? A critical review of Ipswichian (MIS 5e) palaeotemperature reconstructions. *Journal of Quaternary Science* 31, 857–868.

Cannell, M. G. R. & Pitcairn, C. E. R. (1993). *Impacts of the Mild Winters and Hot Summers in the United Kingdom in 1988–1990.* HMSO, London.

Caro, T. (2007). The Pleistocene re-wilding gambit. *Trends in Ecology and Evolution* 22, 281–283.

Catt, J. A. & Henderson, I. F. (1993). Rothamsted Experimental Station: 150 years of agricultural research. The longest continuous scientific experiments? *Interdisciplinary Science Reviews* 18, 365–378.

Cave, A. (2005). *Learning to Breathe.* Hutchinson, London.

Chang, C. C. & Turner, B. J. (2019). Ecological succession in a changing world. *Journal of Ecology* 107, 503–509.

Chantrey, J., Dale, T. D., Read, J. M. et al. (2014). European red squirrel population dynamics driven by squirrelpox at a gray squirrel invasion interface. *Ecology and Evolution* 4, 3788–3799.

Chase, J. M., Abrams, P. A., Grover, J. P. et al. (2002). The interaction between predation and competition: a review and synthesis. *Ecology Letters* 5, 302–315.

Chave, S. P. W. (1984). Duncan of Liverpool – and some lessons for today. *Community Medicine* 6, 61–71.

Christidis, N., McCarthy, M. & Stott, P. A. (2020). The increasing likelihood of temperatures above 30 to 40 °C in the United Kingdom. *Nature Communications* 11, 3093.

Clarke, A. (2017). *Principles of Thermal Ecology.* Oxford University Press, Oxford.

Clements, F. E. (1936). Nature and structure of the climax. *Journal of Ecology* 24, 252–284.

Cohen, J. E. (1995). *How Many People Can the Earth Support?* Norton, New York.

Cohen, J. E. (1998). Cooperation and self-interest: Pareto-inefficiency of Nash equilibria in finite random games. *Proceedings of the National Academy of Sciences of the USA* 95, 9724–9731.

Colman, D. R., Poudel, S., Stamps, B. W., Boyd, E. S. and Spear, J. R. (2017). The deep, hot biosphere: twenty-five years of retrospection. *Proceedings of the National Academy of Sciences of the USA* 114, 6895–6903.

Compton, S. G., Keym R. S. & Key, R. J. D. (2002). Conserving our little Galapagos – Lundy, Lundy Cabbage and its beetles. *British Wildlife* 13, 184–190.

Connell, J. H. (1961a). Effects of competition, predation by *Thais lapillus,* and other factors on natural populations of the barnacle *Balanus balanoides. Ecological Monographs* 31, 61–104.

Connell, J. H. (1961b). The influence of interspecific competition and other factors on the distribution of the barnacle *Chtamalus stellaris. Ecology* 42, 710–723.

Connell, J. H. (1978). Diversity in tropical rain forests and coral reefs. *Science* 199, 1302–1310.

Connell, J. H. (2002). Profiles: Joe Connell. In P. Stiling, *Ecology,* 4th edition. Prentice Hall, Upper Saddle River, pp. 118–119.

Conrad, K. F., Willson, K. H., Harvey, I. F., Thomas, C. J. & Sherratt, T. N. (1999). Dispersal characteristics of seven odonate species in an agricultural landscape. *Ecography* 22, 524–531.

Convey, P., Coulson, S. J., Worland, M. R. & Sjöblom, A. (2018). The importance of understanding annual and shorter-term temperature patterns and variation in the surface levels of polar soils for terrestrial biota. *Polar Biology* 41, 1587–1605.

Conway, V. M. (1936). Studies in the autecology of *Cladium mariscus* R.BR. I. Structure and development. *New Phytologist* 35, 12–204.

Conway, V. M. (1947). Ringinglow Bog near Sheffield: Part I. Historical. *Journal of Ecology* 34, 14–181.

Coulson, J. C. (2011). *The Kittiwake.* T. & A. D. Poyser, London.

Coulson, J. C. & Findlay, I. (2018). Fauna. In S. Gater (ed.), *The Natural History of Upper Teesdale,* 5th edition. Durham Wildlife Trust, Durham.

Coûteaux, M.-M., Bottner, P. & Berg, B. (1995). Litter decomposition, climate and litter quality. *Trends in Ecology and Evolution* 10, 63–66.

Covey, K. R. & Megonigal, J. P. (2019). Methane production and emissions in trees and forests. *New Phytologist* 222, 35–51.

Cramp, S. & Perrins, C. M. (eds.) (1993). *The Birds of the Western Palearctic. Vol VII.* Oxford University Press, Oxford.

Cramp, S. & Perrins, C. M. (eds.) (1994). *The Birds of the Western Palearctic. Vol VIII.* Oxford University Press, Oxford.

Creevy, A. L., Fisher, J., Puppe, D. & Wilkinson, D. M. (2016). Protist diversity on a nature reserve in NW England – with particular reference to their role in soil biogenic silicon pools. *Pedobiologica* 59, 51–59.

Crockatt, M. E. & Bebber, D. P. (2015). Edge effects on moisture reduce wood decomposition rate in a temperate forest. *Global Change Biology* 21, 698–707.

Crowcroft, P. (1991). *Elton's Ecologists.* University of Chicago Press, Chicago.

Csergő, A. M., Demeter, L. & Turkington, R. (2013). Declining diversity in abandoned grasslands of the Carpathian Mountains: do dominant species matter? *PLoS ONE* 8(8), e73533. doi:10.1371/journal. pone.0073533.

Daniels, R. E. (1989). Adaptation and variation in bog mosses. *Plants Today* 2, 139–144.

Darwin, C. (1842). Notes on the Effects produced by the Ancient Glaciers of

Caernarvonshire, and on the Boulders transported by Floating Ice. *The London, Edinburgh and Dublin Philosophical Magazine* 21 (September), 180–188.

Darwin, C. (1859). *On the Origin of Species by Means of Natural Selection.* John Murray, London.

Darwin, C. (1875). *Insectivorous Plants.* John Murray, London.

Darwin, C. (1881). *The Formation of Vegetable Mould Through the Action of Worms, with Observations on their Habits.* John Murray, London.

Davies, N. B., Krebs, J. R. & West, S. A. (2012). *An Introduction to Behavioural Ecology*, 4th edition. Wiley-Blackwell, Chichester.

Dawkins, R. (1976). *The Selfish Gene.* Oxford University Press, Oxford.

Dawkins, R. (1982). *The Extended Phenotype.* Oxford University Press, Oxford.

de Beer, G. (1974). *Charles Darwin and Thomas Henry Huxley. Autobiographies.* Oxford University Press, London.

Defoe, D. (1722). *A Journal of the Plague Year.* E. Nutt and others, London.

Djukic, I., Kepfer-Rojas, S., Schmidt, I. K. et al. (2018). Early stage litter decomposition across biomes. *Science of the Total Environment* 628–629, 1369–1394.

Doolittle, W. F. (1981). Is nature really motherly? *Coevolution Quarterly* 29, 58–63.

Doolittle, W. F. (2014). Natural selection through survival alone, and the possibility of Gaia. *Biology and Philosophy* 29, 415–423.

Doolittle, W. F. (2017). Darwinizing Gaia. *Journal of Theoretical Biology* 434, 11–19.

Downing, K. & Zvirinsky, P. (1999). The simulated evolution of biochemical guilds: reconciling Gaia theory and natural selection. *Artificial Life* 5, 291–318.

Egerton, F. N. (2012). *Roots of Ecology.* University of California Press, Berkeley.

Ellis, C. J., Crittenden, P. D., Scrimgeour, C. M. & Ashcroft, C. J. (2005). Translocation of ^{15}N indicates nitrogen recycling in the mat-forming lichen *Cladonia portentosa. New Phytologist* 168, 423–434.

Ellis, C. J., Yahr, R. & Coppins, B. J. (2011). Archaeological evidence for a massive loss of epiphyte species richness during industrialization in southern England. *Proceedings of the Royal Society B* 278, 3482–3489.

Ellison, A. M. & Gotelli, N. J. (2009). Energetics and the evolution of carnivorous plants: Darwin's 'most wonderful plants in the world'. *Journal of Experimental Botany* 60, 19–42.

Elton, C. (1927). *Animal Ecology.* Sidgwick and Jackson, London.

Ennion, E. A. R. (1949). *Adventurers Fen*, 2nd edition. Herbert Jenkins, London.

Fara, P. (2009). A microscopic reality tale. *Nature* 459, 642–644.

Fenn, K., Malhi, Y., Morecroft, M., Lloyd, C. & Thomas, M. (2015). The carbon cycle of a maritime ancient temperate broadleaved woodland at seasonal and annual scales. *Ecosystems* 18, 1–15.

Fielding, J. J., Croudace, I. W., Kemp, A. E. S. et al. (2020). Tracing lake pollution, eutrophication and partial recovery from the sediments of Windermere, UK, using geochemistry and sediment microfabrics. *Science of the Total Environment* 722, 137745.

Fisher, J., Barker, T., James, C. & Clarke, S. (2009). Water quality in chronically nutrient-rich lakes: the example of the Shropshire–Cheshire meres. *Freshwater Reviews* 2, 79–99.

Fisher, N., Brown, N. D. & Savill, P. S. (2010). Conservation and management of Wytham Woods. In P. S. Savill, C. M. Perrins, K. J. Kirby & N. Fisher (eds.), *Wytham Woods, Oxford's Ecological Laboratory.* Oxford University Press, Oxford. pp. 197–215.

Fitter, R. S. R. (1945). *London's Natural History.* New Naturalist 3. Collins, London.

Flemming, A. (2008). Bears in Britain. *British Wildlife* 19, 393–398.

Foissner, W. & Wanner, M. (1995). Protozoa of soil. In G. Brugerolle & J.-P. Mingot (eds.), *Protistological Actualities.* Clermont-Ferrand, pp. 128–135.

Forbes, S. A. (1887). The lake as a microcosm. *Bulletin of the Peoria Scientific Association* 1887, 77–87. (Reprinted in *Bulletin of the Illinois State Natural History Survey* (1925) 15, 537–550).

Foster, D. R. (1992). Land-use history (1730–1990) and vegetation dynamics in central New England. *Journal of Ecology* 80, 753–772.

Fox, J. W. (2013). The intermediate disturbance hypothesis should be abandoned. *Trends in Ecology and Evolution* 28, 86–92.

Friday, L. E. (1991). The size and shape of traps of *Utricularia vulgaris*. *Functional Ecology* 5, 602–607.

Friday, L. E. (ed.) (1997). *Wicken Fen: the Making of a Wetland Reserve.* Harley Books, Colchester.

Friday, L. E., Grubb, P. J. & Coombe, D. E. (1999). The Godwin Plots at Wicken Fen: a 55-year record of the effects of mowing on fen vegetation. *Nature in Cambridgeshire* 41, 33–46.

Fyfe, R. (2007). The importance of local-scale openness within regions dominated by closed woodland. *Journal of Quaternary Science* 22, 571–578.

Gallagher, M. C., Davenport, J., Gregory, S. et al. (2013) The invasive barnacle species, *Austrominius modestus*: its status and competition with indigenous barnacles on the Isle of Cumbrae, Scotland. *Estuarine, Coastal and Shelf Science* 152, 134–141.

Gao, Z., Karlsson, I., Geisen, S., Kowalchuk, G. & Jousset, A. (2019). Protists: puppet masters of the rhizosphere microbiome. *Trends in Plant Science* 24, 165–176.

Gause, G. (1934). *The Struggle for Existence.* Williams & Williams, Baltimore.

Gest, H. (2004). The discovery of microorganisms by Robert Hooke and Antoni van Leeuwenhoek, Fellows of the Royal Society. *Notes and Records of the Royal Society* 8, 187–201.

Gibb, J. (1954). Feeding ecology of tits, with notes on Treecreepers and Goldcrests. *Ibis* 96, 513–543.

Gilbert, O. (2000). *Lichens.* New Naturalist 86. Collins, London.

Godfray, H. C. J., Aveyard, P., Garnett, T. et al. (2018). Meat consumption, health, and the environment. *Science* 361, eaam5324.

Godwin, H. (1978). *Fenland: its Ancient Past and Uncertain Future.* Cambridge University Press, Cambridge.

Godwin, H., Clowes, D. R. & Huntley, B. (1974). Studies in the ecology of Wicken Fen. V. Development of fen carr. *Journal of Ecology* 62, 197–214.

Goode, D. (2014). *Nature in Towns and Cities.* New Naturalist 127. Collins, London.

Gould, S. J. (1981). *The Mismeasure of Man.* W.W. Norton, New York.

Gould, S. J. (2002). *The Structure of Evolutionary Theory.* Belknap Press, Cambridge, MA.

Grace, J. (2019). Has ecology grown up? *Plant Ecology and Diversity* 12, 387–405.

Green, S., Elliot, M., Armstrong, A. & Hendry, S. J. (2015). *Phytophthora austrocedrae* emerges as a serious threat to juniper (*Juniperus communis*) in Britain. *Plant Pathology* 64, 456–466.

Grenfell, B. & Keeling, M. (2007). Dynamics of infectious disease. In R. M. May & A. R. McLean (eds.),*Theoretical Ecology*, 3rd edition. Oxford University Press, Oxford, pp. 132–147.

Griffiths, J. W. & Henfrey, A. (1875). *Micrographic Dictionary*, 3rd edition. John van Voorst, London.

Grime, J. P. (1998). Benefits of plant diversity to ecosystems: intermediate, filter and founder effects. *Journal of Ecology* 86, 902–910.

Grime, J. P. (2001). *Plant Strategies, Vegetation Processes, and Ecosystem Properties*, 2nd edition. John Wiley & Sons, Chichester.

Grime, J. P. & Blythe, G. M. (1969). An investigation of the relationship between snails and vegetation at the Winnats Pass. *Journal of Ecology* 57, 45–66.

Grime, J. P., MacPherson-Stewart, S. F. & Dearman, R. S. (1968). An investigation of leaf palatability using the snail *Cepaea nemoralis* L. *Journal of Ecology* 56, 405–420.

Grime, J. P., Mackay, J. M. L., Hiller, S. M. & Reid, D. J. (1987). Floristic diversity in a model system using experimental microcosms. *Nature* 328, 420–422.

Grime, J. P., Hodgson, J. G. & Hunt, R. (2007). *Comparative Plant Ecology; a Functional Approach to Common British Species*, 2nd edition. Castlepoint Press, Dalbeattie.

Grodwohl, J.-B., Porto, F. & El-Hani, C. N. (2018). The instability of field experiments: building an experimental research tradition on the rock shore (1950–1985). *History and Philosophy of the Life Sciences* 40, 45.

Groom, D. W. (1993). Magpie *Pica pica* predation on blackbird *Turdus merula* nests in urban areas. *Bird Study* 40, 55–62.

Grubb, P. J. (2016). Trade-offs in interspecific comparisons in plant ecology and how plants overcome proposed constraints. *Plant Ecology and Diversity* 9, 3–33.

Grube, M. & Berg, G. (2009). Microbial consortia of bacteria and fungi with focus on the lichen symbiosis. *Fungal Biology Reviews* 23, 72–85.

Guerrero, R., Margulis, L. & Berlanga, M. (2013). Symbiogenesis: the holobiont as a unit of evolution. *International Microbiology* 16, 133–143.

Halliday, G. (1997). *A Flora of Cumbria*. University of Lancaster, Lancaster.

Hambler, C., Wint, G. R. W. & Rodgers, D. J. (2010). Invertebrates. In P. S. Savill, C. M. Perrins, K. J. Kirby & N. Fisher (eds.), *Wytham Woods, Oxford's Ecological Laboratory*. Oxford University Press, Oxford, pp. 109–144.

Hambler, C., Henderson, P. A. & Speight, M. R. (2011). Extinction rates, extinction-prone habitats, and indicator groups in Britain and at larger scales. *Biological Conservation* 144, 713–721.

Hanski, I. (1983). Coexistence of competitors in patchy environments. *Ecology* 64, 493–500.

Hanski, I. (1998). Metapopulation dynamics. *Nature* 396, 41–49.

Hanski, I. (2016). *Message from Islands*. University of Chicago Press, Chicago.

Hardin, G. (1960). The competitive exclusion principle. *Science* 131, 1292–1297.

Harmer, R., Peterken, G., Kerr, G. & Poulton, P. (2001). Vegetation changes during 100 years of development of two secondary woodlands on abandoned arable land. *Biological Conservation* 101, 291–304.

Harris, J. A., Hobbs, R. J., Higgs, E. & Aronson, J. (2006). Ecological restoration and global climate change. *Restoration Ecology* 14, 170–176.

Hauck, M., Otto, P. I., Dittrich, S. *et al.* (2011). Small increase in sub-stratum pH causes the dieback of one of Europe's most common lichens *Lecanoar conizaeoides*. *Annals of Botany* 108, 359–366.

Hawksworth, D. L. & Grube, M. (2020). Lichens redefined as complex ecosystems. *New Phytologist* 227, 1281–1283.

Hawksworth, D. L. & Seaward, M. R. D. (1977). *Lichenology in the British Isles 1568–1975*. Richmond Publishing Co., Richmond.

Hayhow, D. B., Johnstone, I., Moore, A. S. *et al.* (2018). Breeding status of red-billed choughs *Pyrrhocorax pyrrhocorax* in the UK and Isle of Man in 2014. *Bird Study* 65, 458–470.

Heal, O. W. (1962). The abundance and micro-distribution of testate amoebae (Rhizopoda: Tesacea) in *Sphagnum*. *Oikos* 13, 35–47.

Heal, O. W. (1964). Observations on the seasonal and spatial distribution of testacea (Protozoa: Rhizopoda) in *Sphagnum*. *Journal of Animal Ecology* 33, 395–412.

Heald, O. J. N., Fraticelli, C., Cox, S. E. *et al.* (2020). Understanding the origin of the ring-necked parakeet in the UK. *Journal of Zoology* 312, 1–11.

Heger, T. J., Mitchell, E. A. D. & Leander, B. S. (2013). Holarctic phylogeography of the testate amoeba *Hyalophenia papilio* (Amoebozoa: Arcellinida) reveals extensive genetic diversity explained more by environment than dispersal. *Molecular Ecology* 22, 5172–5184.

Helgason, T., Daniell, T. J., Husband, D. R., Fitter, A. H. & Young, J. P. W. (1998). Ploughing up the wood-wide web. *Nature* 394, 431.

Hibbett, D., Blanchette, R., Kenrick, P. & Mills, B. (2016). Climate decay, and the death of coal forests. *Current Biology* 26, R543–R576.

Hicks, S. P. (1971). Pollen-analytical evidence for the effect of prehistoric agriculture on the vegetation of north Derbyshire. *New Phytologist* 70, 647–667.

Hill, M. O., Evans, D. F. & Bell, S. A. (1992). Long-term effects of excluding sheep from hill pastures in North Wales. *Journal of Ecology* 80, 1–13.

Hills, J. M. & Thomason, C. J. (2002). The 'ghost of settlement past' determines mortality and fecundity in the barnacle, *Semibalanus balanoides*. *Oikos* 101, 529–538.

Ho, J. C., Michalak, A. M. & Pahlevan, N. (2019). Widespread global increase in intense lake phytoplankton blooms since the 1980s. *Nature* 574, 667–670.

Hobbs, R. J., Higgs, E. S. & Hall, C. M. (2013). *Novel Ecosystems: Intervening in the New World Order*. Wiley-Blackwell, Chichester.

Holloway, S. (1996). *The Historical Atlas of Breeding Birds in Britain and Ireland, 1875–1900*. T. & A. D. Poyser, London.

Holmes, E. C. (2011). Plague's progress. *Nature* 478, 465–466.

Honegger, R., Edwards, D. & Axe, L. (2013). The earliest records of internally stratified cyanobacterial and algal lichens from the Lower Devonian of the Welsh Borderland. *New Phytologist* 197, 264–275.

Hoopes, M. F. & Harrison, S. (1998). Metapopulation, source–sink and disturbance dynamics. In W. J. Sutherland (ed.), *Conservation, Science and Action*. Blackwell, Oxford, pp. 135–151.

Hubbell, S. P. (2001). *The Unified Neutral Theory of Biodiversity and Biodiversity*. Princeton University Press, Princeton.

Hug, L., Baker, B. J., Anantharaman, K. *et al.* (2016). A new view of the tree of life. *Nature Microbiology* 1, 16048.

Huggett, R. J. (1999). Ecosphere, biosphere or Gaia? What to call the global ecosystem. *Global Ecology and Biogeography* 8, 425–431.

Huneman, P. (2018). Stephen Hubbell and the paramount power of randomness. In O. Harman & M. R. Dietrich (eds.), *Dreamers, Visionaries, and Revolutions in the Life Sciences*. Chicago University Press, Chicago, pp. 176–195.

Huston, M. A. (2014). Disturbance, productivity, and species diversity: empiricism vs. logic in ecological theory. *Ecology* 95, 2382–2396.

Hutchinson, G. E. (1957). Concluding remarks. *Cold Spring Harbor Symposia on Quantitative Biology* 22, 415–427.

Hutchinson, G. E. (1961). The paradox of the plankton. *American Naturalist* 95, 137–145.

Hutchinson, G. E. (1964). The lacustrine microcosm reconsidered. *American Scientist* 52, 334–341.

Hutchinson, G. E. (1965). *The Ecological Theatre and the Evolutionary Play*. Yale University Press, New Haven.

Janzen, D. H. (1977). Why fruits rot, seeds mold, and meat spoils. *American Naturalist* 111, 691–713.

Jenkins, D. G. (2014). Lakes and rivers as microcosms, version 2.0. *Journal of Limnology* 73, 20–32.

Johnson, E. A. & Miyanishi, K. (2008). Testing the assumptions of chronosequences in succession. *Ecology Letters* 11, 419–431.

Johnston, A. E. & Poulton, P. R. (2018). The importance of long-term experiments in agriculture: their management to ensure continued crop production and soil fertility: the Rothamsted experience. *European Journal of Soil Science* 69, 113–125.

Jones, A. R. (1974). *The Ciliates*. Hutchinson, London.

Jones, C. G., Lawton, J. H. & Shachak, M. (1994). Organisms as ecosystem engineers. *Oikos* 69, 373–386.

Jones, C. G., Lawton, J. H. & Shachak, M. (1997). Positive and negative effects of organisms as physical ecosystem engineers. *Ecology* 78, 1946–1957.

Jones, C. G., Gutiérrez, J. L., Byers, J. E. *et al.* (2010). A framework for understanding physical ecosystem engineering by organisms. *Oikos* 119, 1826–1869.

Jones, J. M. (1987). Chemical fractionation of copper, lead and zinc in ombrotrophic peat. *Environmental Pollution* 48, 131–144.

Jones, R. (2018). *Beetles*. New Naturalist 136. William Collins, London.

Kahan, D. M. (2017). 'Ordinary science intelligence': a science-comprehension measure for study of risk and science communication, with notes on evolution and climate change. *Journal of Risk Research* 20, 995–1016.

Keddy, P. (2005). Putting the plants back into plant ecology; six pragmatic models for understanding and conserving plant diversity. *Annals of Botany* 96, 177–189.

Keddy, P. A. (2007). *Plants and Vegetation*. Cambridge University Press, Cambridge.

Keeton, W. T. (1980). *Biological Science*, 3rd edition. W. W. Norton, New York.

Kenward, R. E., Hodder, K. H., Rose, R. J. *et al.* (1998). Comparative demography of red squirrels (*Sciurus vulgaris*) and grey squirrels (*Sciurus carolinensis*) in deciduous and conifer woodland. *Journal of Zoology* 244, 7–21.

Kibby, G. (2017). *Mushrooms and Toadstools of Britain and Europe.* Vol 1. Self-published.

Kingsland, S. E. (1995). *Modelling Nature*, 2nd edition. University of Chicago Press, Chicago.

Kington, J. (2010). *Climate and Weather.* New Naturalist 115. Collins, London.

Kirby, K. (2016). Charles Elton and Wytham Woods. *British Wildlife* 27, 256–263.

Kleidon, A. (2016). *Thermodynamic Foundations of the Earth System.* Cambridge University Press, Cambridge.

Klein, T., Siegwolf, R. T. W. & Körner, K. (2016). Belowground carbon trade amongst tall trees in a temperate forest. *Science* 352, 342–344.

Klinger, L. F. (1996). The myth of the classic hydrosere model of bog succession. *Arctic and Alpine Research* 28, 1–9.

Knowles, R. D., Pastor, J. & Biesboer, D. D. (2006). Increased soil nitrogen associated with dinitrogen-fixing, terricolous lichens of the genus *Peltigera* in northern Minnesota. *Oikos* 114, 37–48.

Kokko, H. (2017). Give one species the task to come up with a theory that spans them all: what good can come from that? *Proceedings of the Royal Society B.* 284, 20171652.

Koumoutsaris, S. (2019). A hazard model of sub-freezing temperatures in the United Kingdom using vine copulas. *Natural Hazards and Earth Systems Science* 19, 489–506.

Krebs, C. J. (2009). *Ecology*, 6th edition. Benjamin Cummings, San Francisco.

Krebs, J. (2010). Preface. In P. S. Savill, C. M. Perrins, K. J. Kirby & N. Fisher (eds.), *Wytham Woods, Oxford's Ecological Laboratory.* Oxford University Press, Oxford, pp. ix–x.

Kreyling, J., Schweiger, A. H., Bahn, M. *et al.* (2018). To replicate, or not to

replicate – that is the question: how to tackle nonlinear responses in ecological experiments. *Ecology Letters* 21, 1629–1638.

Kruuk, H. (2003). *Niko's Nature.* Oxford University Press, Oxford.

Lack, D. (1946). *The Life of the Robin*, revised edition. Witherby, London.

Lack, D. (1971). *Ecological Isolation in Birds.* Blackwell, Oxford.

Lamb, H. H. (1982). *Climate, History and the Modern World.* Methuen, London.

Lamentowicz, M., Bragazza, L., Buttler, A., Jassey, V. E. J. & Mitchell E. A. D. (2013). Seasonal patterns of testate amoebae diversity, community structure and species–environment relationship in four *Sphagnum*-dominated peatlands along a 1300 m altitudinal gradient in Switzerland. *Soil Biology and Biochemistry* 67, 1–11.

Lamm, E. (2018). Big dreams for small creatures. In O. Harman & M. R. Dietrich (eds.), *Dreamers, Visionaries, and Revolutions in the Life Sciences.* Chicago University Press, Chicago, pp. 288–304.

Lang, T. (2020). *Feeding Britain.* Pelican, London.

Langley, P. J. W. & Yalden, D. W. (1977). The decline of the rarer carnivores in Great Britain during the nineteenth century. *Mammal Review* 7, 95–116.

Lara, E. & Gomaa, F. (2017). Symbiosis between testate amoebae and photosynthetic organisms. In M. Grube, J. Seckbach & L. Muggia (eds.), *Algal and Cyanobacterial Symbioses.* World Scientific, pp. 399–420.

Lara, E., Dunmack, K., García-Martín, J. M., Kudryavtsev, A. & Kosakyan, A. (2020). Amoeboid protist systematics: a report on the 'Systematics of amoeboid protists' symposium at the VIIIth ECOP/ISOP meeting, Rome, 2019. *European Journal of Protistology* 76, 125727.

Latour, B. (1999). *Pandora's Hope.* Harvard University Press, Cambridge, MA.

Lawton, J. H. (1996). Population abundances, geographic ranges and conservation. 1994 Witherby Lecture. *Bird Study* 43, 3–19.

Lawton J. H. (1999a). Are there general laws in ecology? *Oikos* 84, 177–192.

Lawton, J. H. (1999b). Time to reflect: Gilbert White and environmental change. *Oikos* 87, 476–478.

Lawton, J. H. (2001). Earth systems science. *Science* 292, 1965.

Lawton, J. H. & May, R. M. (1983). The birds of Selborne. *Nature* 306, 732–733.

Leidy, J. (1879). *Fresh-Water Rhizopods of North America*. United States Geological Survey, Washington.

Lenton, T. (2016). *Earth System Science: a Very Short Introduction*. Oxford University Press, Oxford.

Lenton, T. M. & Latour, B. (2018). Gaia 2.0. *Science* 361, 1066–1068.

Lenton, T. M. & Vaughan, N. E. (2009). The radiative forcing of different climate geoengineering options. *Atmospheric Chemistry and Physics* 9, 5539–5561.

Lenton, T. M. & Watson, A. (2011). *Revolutions that Made the Earth*. Oxford University Press, Oxford.

Lenton, T. M., Daines, S. J., Dyke, J. G. *et al.* (2018). Selection for Gaia across multiple scales. *Trends in Ecology and Evolution* 33, 633–645.

Leuzinger, S. & Körner, C. (2007). Tree species diversity affects canopy leaf temperatures in a mature temperate forest. *Agricultural and Forest Meteorology* 146, 29–37.

Lever, C. (1977). *The Naturalised Animals of the British Isles*. Hutchinson, London.

Levins, R. (1970). Extinction. In M. Gesterhaber (ed.), *Some Mathematical Problems in Biology*. American Mathematical Society, Providence, pp. 77–107.

Lewis, S., Sherratt, T. N., Hamer, K. C. & Wanless, S. (2001). Evidence of intra-specific competition for food in a pelagic seabird. *Nature* 412, 816–819.

Lindeman, R. L. (1942). The trophic-dynamic aspect of ecology. *Ecology* 23, 399–418.

Lineweaver, C. H. & Egan, C. A. (2008). Life, gravity and the second law of thermodynamics. *Physics of Life Reviews* 5, 225–242.

Liu, L., Chen, H., Yang, J. R. *et al.* (2019). Response of the eukaryotic plankton community to the cyanobacterial biomass cycle over 6 years in two subtropical reservoirs. *ISME Journal* 13, 2196–2208.

Loarie, S. R., Duffy, P. B., Hamilton, H. *et al.* (2009). The velocity of climate change. *Nature* 462, 1052–1055.

Lovelock, J. (1995). *The Ages of Gaia*, 2nd edition. Oxford University Press, Oxford.

Lovelock, J. (2003). The living Earth. *Nature* 426, 769–770.

Lovelock, J. (2006). *The Revenge of Gaia*. Allen Lane, London.

Loveluck, C. P., More, A. F., Spaulding, N. E. *et al.* (2020). Alpine ice and the annual political economy of the Angevin Empire, from the death of Thomas Becket to Magna Carta, c. AD 1170–1216. *Antiquity* 94, 473–490.

Lovett, C. (2006). *Agrocybe rivulosa* new to the British list. *Field Mycology* 7, 47–48.

Lowe, J. A., Bernie, D., Bett, P. *et al.* (2018, updated 2019). *UKCP18 Science Overview Report*. Met Office, Exeter.

Maberly, S. C. & Elliott, J. A. (2012). Insights from long-term studies in the Windermere catchment: external stressors, internal interactions and the structure and function of lake ecosystems. *Freshwater Biology* 57, 233–243.

Maberly, S. C., Ciar, D., Elliott, J. A. *et al.* (2018). From ecological informatics to the generation of ecological knowledge: long-term research in the English Lake District. In F. Recknagel & W. K. Michener (eds.), *Ecological informatics.*, 3rd edition. Springer, Cham, pp. 455–482.

Mabey, R. (1986). *Gilbert White*. Century, London.

Mabey, R. (1996). *Flora Britannica*. Sinclair-Stevenson, London.

Macan, T. T. (1970). *Biological Studies of the English Lakes*. Longman, London.

MacArthur, R. H. (1958). Population ecology of some warblers of northeastern coniferous forests. *Ecology* 39, 599–619.

Macdonald, D. W., Newman, C., Dean, J., Buesching, C. D. & Johnson, P. J. (2004). The distribution of Eurasian badger, *Meles meles*, setts in a high-density area: field observations contradict the sett dispersion hypothesis. *Oikos* 106, 295–307.

Mackey, R. L. and Currie, D. J. (2001). The diversity–disturbance relationship: is it generally strong and peaked? *Ecology* 82, 3479–3492.

Madigan, M., Martinko, J., Stahl, D. & Clark, D. (2012). *Brock: Biology of Microorganisms,* 13th edition. Pearson, Boston.

Marchant, J. H., Freeman, S. N., Crick, H. Q. P. & Beaven, L. P. (2004). The BTO heronries census of England and Wales 1928–2000: new indices and a comparison of analytical methods. *Ibis* 146, 323–334.

Margulis, L. & Chapman, M. J. (2009). *Kingdoms and Domains.* Academic Press, London.

Margulis, L. & Lovelock, J. E. (1974). Biological modulation of the Earth's atmosphere. *Icarus* 21, 471–489.

Margulis, L. & Sagan, D. (1995). *What is Life?* Weidenfeld & Nicolson, London.

Margulis, L., Dolan, M. F. & Guerrero, R. (2000). The chimeric eukaryote: origin of the nucleus from the karyomastigont in amitochondriate protists. *Proceedings of the National Academy of Sciences of the USA* 97, 6954–6959.

Marren, P. (1999). *Britain's Rare Flowers.* Poyser, London.

Marren, P. (2016). The great rewilding experiment at Knepp Castle. *British Wildlife* 27, 333–399.

Martin, D., Thompson, A., Stewart, I. *et al.* (2012). A paradigm of fragile Earth in Priestley's bell jar. *Extreme Physiology and Medicine* 1, 4.

Massimino, D., Woodward, I. D., Hammond, M. J. *et al.* (2019). BirdTrends 2019: trends in numbers, breeding success and survival for UK breeding birds. Research Report 722. BTO, Thetford. www.bto.org/birdtrends (accessed October 2020).

Matless, D. (2016). *Landscape and Englishness,* 2nd edition. Reaktion Books, London.

May, R. M., Crawley, M. J. & Sugihara, G. (2007). Communities: patterns. In R. M. May & A. R. McLean (eds.), *Theoretical Ecology,* 3rd edition. Oxford University Press, Oxford, pp. 111–131.

Mayr, E. (1991). *One Long Argument.* Harvard University Press, Cambridge, MA.

McGovern, S. T., Evans, C. D., Dennis, P. *et al.* (2014). Increased inorganic nitrogen leaching from a mountain grassland ecosystem following grazing removal: a hangover of past intensive land use? *Biogeochemistry* 119, 125–138.

McGuffie, K. & Henderson-Sellers, A. (2005). *A Climate Modelling Primer,* 3rd edition. John Wiley & Sons, Chichester.

McHaffie, H. S., Legg, C. J., Proctor, M. C. F. & Amphlett, A. (2009). Vegetation and ombrotrophy on the mires of Abernethy Forest, Scotland. *Plant Ecology and Diversity* 2, 5–103.

McLean, A. R. & May, R. M. (2007). Introduction. In R. M. May & A. R. McLean (eds.), *Theoretical Ecology,* 3rd edition. Oxford University Press, Oxford, pp. 1–6.

Medawar, P. (1982). *Pluto's Republic.* Oxford University Press, Oxford.

Merryweather, J. & Fitter, A. (1995). Phosphorus and carbon budgets: mycorrhizal contribution in *Hyacinthoides non-scripta* (L.) Chouard ex Rothm under natural conditions. *New Phytologist* 129, 619–627.

Merryweather, J. & Fitter, A. (1998). The arbuscular mycorrhizal fungi of *Hacinthoides non-scripta*. I. Diversity of fungal taxa. *New Phytologist* 138, 117–129.

Midgley, M. (2001). *Science and Poetry.* Routledge, London.

Milcu, A. & Manning, P. (2011). All size classes of soil fauna and litter quality control the acceleration of litter decay in its home environment. *Oikos* 120, 1366–1370.

Mitchell, R. J., Hewison, R. L., Hester, A. J., Broome, A. & Kirby, K. J. (2016). Potential impacts of the loss of *Fraxinus excelsior* (Oleaceae) due to ash dieback on woodland vegetation in Great Britain. *New Journal of Botany* 6, 2–15.

Moore, O. & Crawley, M. J. (2015). The impact of red deer management on cryptogam ecology in vegetation typical of north west Scotland. *Plant Ecology and Diversity* 8, 127–137.

Moore, P. D. (2001). A never ending story. *Nature* 409, 565.

Morris, B. E. L., Henneberger, R., Huber, H. & Moissl-Eichinger, C. (2013). Microbial syntrophy: interaction for the common good. *FEMS Microbiology Reviews* 37, 384–406.

Morton, O. (2007). *Eating the Sun: How Plants Power the Planet.* Fourth Estate, London.

Moss, B. (1989). Water pollution and the management of ecosystems: a case study of science and scientists. In P. J. Grubb & J. B. Whittaker (eds.), *Towards a More Exact Ecology*. Blackwell, Oxford, pp. 401–422.

Moss, B. (2001). *The Broads*. New Naturalist 89. Collins, London.

Moss, B. (2010). *Ecology of Freshwaters*, 4th edition. Wiley-Blackwell, Chichester.

Moss, B. (2015). *Lakes, Loughs and Lochs*. New Naturalist 128. Collins, London.

Moss, B. (2017). *Ponds and Small Lakes: Microorganisms and Freshwater Ecology*. Naturalists' Handbook No 32. Pelagic Publishing, Exeter.

Moss, C. E. (1913). *Vegetation of the Peak District*. Cambridge University Press, Cambridge.

Munzi, S., Correia, O., Silva, P. *et al.* (2014). Lichens as ecological indicators in urban areas: beyond the effects of pollutants. *Journal of Applied Ecology* 51, 1750–1757.

Nagy, L. & Grabherr, G. (2009). *The Biology of Alpine Habitats*. Oxford University Press, Oxford.

Nauta, M. M. (2003). A new *Agrocybe* on woodchip in North Western Europe. *Persoonia* 18, 271–274.

Nee, S. (2007). Metapopulations and their dynamics. In R. M. May & A. R. McLean (eds.), *Theoretical Ecology*, 3rd edition. Oxford University Press, Oxford, pp. 35–45.

Nelsen, M. P., DiMichele, W. A., Peters, S. E. & Boyce, C. K. (2016). Delayed fungal evolution did not cause the Paleozoic peak in coal production. *Proceedings National Academy Science USA* 113, 2442–2447.

Nelson, B. (2002). *The Atlantic Gannet*, 2nd edition. Fenix Books, Great Yarmouth.

Newman, E. I. (1988). Mycorrhizal links between plants: Their functioning and ecological significance. *Advances in Ecological Research* 18, 243–270.

Newsham, K. K., Fitter, A. H. & Watkinson, A. R. (1995). Multi-functionality and biodiversity in arbuscular mycorrhizae. *Trends in Ecology and Evolution* 10, 407–411.

Newton, I. (1992). Experiments on the limitation of bird numbers by territorial behaviour. *Biological Reviews* 67, 129–173.

Newton, I. (2013). *Bird Populations*. New Naturalist 124. William Collins, London.

Newton, I. (2017). *Farming and Birds*. New Naturalist 135. William Collins, London.

Nicholson, A. E., Wilkinson, D. M., Williams, H. T. P. & Lenton, T. M. (2018). Alternative mechanisms for Gaia. *Journal of Theoretical Biology* 457, 249–257.

Norris, T. (2020). Swifts in Selborne. Hampshire Swifts website. www.hampshireswifts.co.uk/selborne (accessed October 2020).

North, F. J., Campbell, B. & Scott, R. (1949). *Snowdonia: the National Park of North Wales*. New Naturalist 13. Collins, London.

Nylander, W. (1866). Les lichens du Jardin du Luxembourg. *Bulletin du Société Botanique de France* 13: 365–372.

O'Connor, R. J. (1982). Habitat occupancy and the regulation of clutch size in the European Kestrel *Falco tinnunculus*. *Bird Study* 29, 17–26.

Odenbaugh, J. (2013). Searching for patterns, hunting for causes. Robert MacArthur, the mathematical naturalist. In O. Harman & M. R. Dietrich (eds.), *Outsider Scientists: Routes to Innovation in Biology*. University of Chicago Press, Chicago, pp. 181–198.

Odling-Smee, F. J., Laland, K. N. & Feldman, M. W. (2003). *Niche Construction*. Princeton University Press, Princeton.

Oliver, R. L. & Ganf, G. G. (2000). Freshwater blooms. In B. A. Whitton & M. Potts (eds.), *The Ecology of Cyanobacteria*. Kluwer Academic, Dordrecht, pp. 149–194.

Ollerton, J., Erenler, H., Edwards, M. & Crockett, R. (2014). Extinctions of aculeate pollinators in Britain and the role of large-scale agricultural changes. *Science* 346, 1360–1362.

O'Malley, M. A. (2017). From endosymbiosis to holobionts: evaluating a conceptual legacy. *Journal of Theoretical Biology* 434, 34–41.

Parker, G. A. (1970). Sperm competition and its evolutionary consequences in insects. *Biological Reviews* 45, 525–567.

Patenaude, P. L., Briggs, B. D. J., Milne, R. *et al.* (2003). The carbon pool in a British semi-natural woodland. *Forestry* 75, 109–119.

Pearsall, W. H. (1950). *Mountains and Moorlands.* New Naturalist 11. Collins, London.

Perrins, C. M. (1979). *British Tits.* New Naturalist 62. Collins, London.

Perrins, C. M. and Gosler, A. G. (2010). Birds. In P. S. Savill, C. M. Perrins, K. J. Kirby & N. Fisher (eds.), *Wytham Woods, Oxford's Ecological Laboratory.* Oxford University Press, Oxford. pp. 145–171.

Pescott, O. L., Simkin, J. M., August, T. A. *et al.* (2015). Air pollution and its effects on lichens, bryophytes, and lichen feeding moths: review and evidence from biological records. *Biological Journal of the Linnean Society* 115, 611–635.

Peterken, G. (2013). *Meadows.* British Wildlife Publishing, Oxford.

Phillips, G., Willby, N. & Moss, B. (2016). Submerged macrophyte decline in shallow lakes: what have we learnt in the last forty years? *Aquatic Botany* 135, 37–45.

Pigott, C. D. (1988). Obituary: Verona Margaret Conway. *Journal of Ecology* 76, 288–291.

Pigott, C. D. & Pigott, S. (1993). Water as a determinant of the distribution of trees at the boundary of the Mediterranean zone. *Journal of Ecology* 81, 557–566.

Pope, D. J. & Lloyd P. S. (1975). Hemispherical photography, topography and plant distribution. In G. C. Evans, R. Bainbridge & O. Rackham (eds.), *Light as an Ecological Factor: II.* Blackwell, Oxford, pp. 385–408.

Porter, R. (1997). *The Greatest Benefit to Mankind.* HarperCollins, London.

Poulton, P. R., Pye, E., Hargreaves, P. R. & Jenkinson, D. S. (2003). Accumulation of carbon and nitrogen by old arable land reverting to woodland. *Global Change Biology* 9, 942–955.

Pozsgai, G., Baird, J., Littlewood, N. A., Pakeman, R. J. & Young, M. R. (2018). Phenological changes of the most commonly sampled ground beetle (Coleoptera: Carabidae) species in the UK environmental change network. *International Journal of Biometeorology* 62, 1063–1074.

Preece, R. C. & Day, S. P. (1994). Comparison of Post-Glacial molluscan and vegetational successions from a radiocarbon-dated tufa sequence in Oxfordshire. *Journal of Biogeography* 21, 463–478.

Preston, C. D., Pearman, D. A. & Dines, T. D. (2002). *New Atlas of the British and Irish Flora.* Oxford University Press, Oxford.

Pritchard, J. O., Porter, A. H. M. & Montagnes, D. J. S. (2016). Did Gause have a yeast infection? *Journal of Eukaryotic Microbiology* 63, 552–557.

Proctor, M. (2013). *Vegetation of Britain and Ireland.* New Naturalist 122. Collins, London.

Pryor, F. (2010). *The Making of the British Landscape.* Allen Lane, London.

Pryor, F. (2019). *The Fens.* Head of Zeus, London.

Pulliam, H. R. (1988). Sources, sinks, and population regulation. *American Naturalist* 132, 652–661.

Pulliam, H. R. (2000). On the relationship between niche and distribution. *Ecology Letters* 3, 349–361.

Quammen, D. (2018). *The Tangled Tree: a Radical New History of Life.* William Collins, London.

Rackham, O. (1986). *The History of the Countryside.* J. M. Dent & Sons, London.

Rackham, O. (1990). *Trees and Woodland in the British Landscape*, 2nd edition. J. M. Dent & Sons, London.

Rackham, O. (2003). *Ancient Woodland*, 2nd edition. Castlepoint Press, Dalbeattie.

Radbourne, A. D., Elliott, J. A., Maberly, S. C., Ryves, D. B. & Anderson, N. J. (2019). The impacts of changing nutrient load and climate on a deep eutrophic, monomictic lake. *Freshwater Biology* 64, 1169–1182.

Raggett, G. F. (1982). Modelling the Eyam plague. *Bulletin of the Institute of Mathematics and its Applications* 18, 221–226.

Ratcliffe, D. A. (1981). The vegetation. In D. Nethersole-Thompson & A. Watson, *The Cairngorms*, revised edition. Melven Press, Perth, pp. 42–76.

Ratcliffe, D. A. (1991). The mountain flora of Britain and Ireland. *British Wildlife* 3, 10–21.

Read, D. J. (2002). Towards ecological relevance: progress and pitfalls in the path towards an understanding of mycorrhizal functions in nature. In

M. G. A. van der Heijden & I. Sanders (eds.), *Mycorrhizal Ecology*. Springer, Berlin, pp. 3–29.

Reczuga, M. K., Lamentowicz, M., Mulot, M. *et al.* (2018). Predator–prey mass ratio drives microbial activity under dry conditions in *Sphagnum* peatlands. *Ecology and Evolution* 8, 5752–5764.

Reynolds, C. (2019). The ecological framework: water chemistry, hydrology and phytoplankton. In T. Wall & G. Wall (eds.), *Rostherne Mere: Aspects of a Wetland Nature Reserve*. Tom Wall, Shropshire, pp. 155–172.

Ricklefs, R. & Relyea, R. (2014). *Ecology: the Economy of Nature*, 7th edition. W. H. Freeman, New York.

Riutta, T., Slade, E. M., Bebber, D. P. *et al.* (2012). Experimental evidence for the interacting effects of forest edge, moisture and soil macrofauna on leaf litter decomposition. *Soil Biology and Biochemistry* 49, 124–131.

Roberts, D. & Zimmer, D. (1990). Microfaunal communities associated with epiphytic lichens in Belfast. *The Lichenologist* 22, 163–171.

Robinson, J. M. (1990). Lignin, land plants, and fungi: biological evolution affecting Phanerozoic oxygen balance. *Geology* 15, 607–610.

Romeo, C., McInnes, C. J., Dale, T. D. *et al.* (2019). Disease, invasions and conservation: no evidence of squirrelpox virus in grey squirrels introduced to Italy. *Animal Conservation* 22, 14–23.

Rothschild, M. & Clay, T. (1952). *Fleas, Flukes and Cuckoos*. New Naturalist Monograph 7. Collins, London.

Rousk, J., Brookes, P. C. & Bååth, E. (2011). Fungal and bacterial growth responses to N fertilization and pH in the 150-year 'Park Grass' UK grassland experiment. *FEMS Microbial Ecology* 76, 89–99.

Rowell, T. A. (1997). The history of the fen. In L. Friday (ed.), *Wicken Fen: the Making of a Wetland Reserve*. Harley Books, Colchester, pp. 187–212.

Russell, E. J. (1914). The partial sterilisation of soils. *Nature* 94, 308–311.

Ruxton, G. D. & Humphries, S. (2008). Can ecological and evolutionary arguments solve the riddle of the missing marine insects? *Marine Ecology* 29, 72–75.

Ruxton, G. D., Wilkinson, D. M., Schaefer, H. M. & Sherratt, T. N. (2014). Why fruit rots: theoretical support for Janzen's theory of microbe–macrobe competition. *Proceedings of The Royal Society B* 281, 20133320.

Rydin, H. & Jeglum, J. K. (2013). *The Biology of Peatlands*, 2nd edition. Oxford University Press, Oxford.

Sainsbury, K. A., Shore, R. F., Schofield, H. *et al.* (2019). Recent history, current status, conservation and management of native mammalian carnivore species in Great Britain. *Mammal Review* 49, 171–188.

Salisbury, E. J. (1964a). The origin and early years of the British Ecological Society. *Journal of Animal Ecology* suppl. 33, 13–18 [also published as *Journal of Ecology* suppl. 52, 13–18].

Salisbury, E. J. (1964b). *Weeds and Aliens*, 2nd edition. New Naturalist 43. Collins, London.

Sapp, J. (2009). *The New Foundations of Evolution*. Oxford University Press, Oxford.

Savill, P. S., Perrins, C. M., Kirby, K. J. and Fisher, N. (eds.) (2010). *Wytham Woods, Oxford's Ecological Laboratory*. Oxford University Press, Oxford.

Scheffer, M., Hosper, S. H., Meijer, M.-L., Moss, B. & Jeppesen, E. (1993). Alternative equilibria in shallow lakes. *Trends in Ecology and Evolution* 8, 275–279.

Scheffer, M., Barrett, S., Carpenter, S. R. *et al.* (2015). Creating a safe operating space for iconic ecosystems. *Science* 347, 1317–1319.

Schmidt, O., Dyckmans, J. & Schrader, S. (2016). Photoautotrophic microorganisms as a carbon source for temperate soil invertebrates. *Biology Letters* 12, 20150646.

Schmidt, S., Raven, J. A. & Paungfoo-Lonhienne, C. (2013). The mixotrophic nature of photosynthetic plants. *Functional Plant Biology* 40, 425–438.

Schopf, J. W. (2006). The first billion years: when did life emerge? *Elements* 2, 229–233.

Scott, M. (2016). *Mountain Flowers*. Bloomsbury, London.

Seaward, M. R. D. & Pentecost, A. (2001). Lichen flora of the Malham Tarn area. *Field Studies* 10, 57–92.

Selosse, M.-A. & Roy, M. (2009). Green plants that feed on fungi: facts and questions about mixotrophy. *Trends in Plant Science* 14, 64–70.

Sheehy, E., Sutherland, C., O'Reilly, C. & Lambin, X. (2018). The enemy of my enemy is my friend: native pine marten recovery reverses the decline of red squirrel by suppressing grey squirrel populations. *Proceedings of the Royal Society B.* 285, 20172603.

Shelford, V. E. (1931). Some concepts of bioecology. *Ecology* 12, 455–467.

Sherratt, T. N. & Wilkinson, D. M. (2009). *Big questions in ecology and evolution.* Oxford University Press, Oxford.

Sherratt, T. N., Wilkinson, D. M. & Bain, R. S. (2006). Why fruits rot, seeds mold and meat spoils: a reappraisal. *Ecological Modelling* 192, 618–626.

Shorten, M. (1954). *Squirrels.* New Naturalist Monograph 12. Collins, London.

Shotter, D. (1997). *Romans and Britons in North-West England.* Centre for North West Regional Studies, Lancaster.

Silva, G. G., Green, A. J., Weber, V. *et al.* (2018). Whole angiosperms *Wolffia columbiana* disperse by gut passage through wildfowl in South America. *Biology Letters* 14, 20180703.

Silvertown, J. (1987). Ecological stability: a test case. *American Naturalist* 130, 807–810.

Silvertown, J. (2005). *Demons in Eden: the Paradox of Plant Diversity.* University of Chicago Press, Chicago.

Silvertown, J., Poulton, P., Johnston, E. *et al.* (2006). The Park Grass Experiment 1856–2006: its contribution to ecology. *Journal of Ecology* 94, 801–814.

Skeffington, R., Wilson, E., Maltby, E., Immirzi, P. & Putwain, P. (1998). Acid deposition and blanket mire degradation and restoration. In J. H. Tallis, R. Meade & P. D. Hulme (eds.), *Blanket Mire Degradation.* Macaulay Land Use Research Institute, Aberdeen, pp. 29–37.

Small, R. (2015). Livestock biodiversity: coming of age? *British Wildlife* 27, 17–24.

Smith, D., Whitehouse, N., Bunting, M. J. & Chapman, H. (2010). Can we characterise 'openness' in the Holocene palaeoenvironmental record? Modern analogue studies of insect faunas and pollen spectra from Dunham Massey deer park and Epping Forest, England. *The Holocene* 20, 215–229.

Smith, M. D., Koerner, S. E., Knapp, A. K. *et al.* (2020). Mass ratio effects underlie ecosystem responses to environmental change. *Journal of Ecology* 108, 855–864.

Smith, P. H. (2009). *The Sands of Time Revisited.* Amberley Publishing, Stroud.

Smith, R. S. (1988). Farming and the conservation of traditional meadowland in the Pennine Dales Environmentally Sensitive Area. In M. B. Usher & D. A. B. Thompson (eds.), *Ecological Change in the Uplands.* Blackwell, Oxford, pp. 183–199.

Smith, R. S. & Jones, L. (1991). The phenology of mesotrophic grassland in the Pennine Dales, northern England: historic hay cutting dates, vegetation variation and plant species phenologies. *Journal of Applied Ecology* 28, 42–59.

Smith, R. S., Shield, R. S., Bardgett, R. D. *et al.* (2003). Soil microbial community, fertility, vegetation and diversity as targets in the restoration management of a meadow grassland. *Journal of Applied Ecology* 40, 51–64.

Southwood, T. R. E. & Henderson, P. A. (2000). *Ecological Methods,* 3rd edition. Blackwell Science, Oxford.

Spooner, B. and Roberts, P. (2005). *Fungi.* New Naturalist 96. Collins, London.

Stace, C. (2019). *New Flora of the British Isles,* 4th edition. C&M Floristics, Penrith.

Stace, C. A. & Crawley, M. J. (2015). *Alien Plants.* New Naturalist 129. Collins, London.

Stamp, L. D. (1955). *Man and the Land.* New Naturalist 31. Collins, London.

Steidinger, B. S., Crowther, T. W., Liang, J. *et al.* (2019). Climatic controls of decomposition drive the global biogeography of forest–tree symbioses. *Nature* 569, 404–408.

Stephenson, S. L. (2010). *The Kingdom Fungi.* Timber Press, Portland.

Stevenson, J. H. (1989). Rothamsted: a cradle of agricultural research. *Plants Today* 2, 84–89.

Stewart, I. (2012). *17 Equations that Changed the World*. Profile Books, London.

Stewart, I. & Tall, D. (2015). *The Foundations of Mathematics*, 2nd edition. Oxford University Press, Oxford.

Stopes M. C. (1912). *Botany, or The Modern Study of Plants*. T. C. & E. C. Jack, London.

Summers, R. W. (2018). *Abernethy Forest: the History and Ecology of an Old Scottish Pinewood*. RSPB, Inverness.

Summers, R. W. (2019). Classic wildlife sites: Abernethy Forest. *British Wildlife* 30, 157–165.

Tansley, A. G. (1935). The use and abuse of vegetational concepts and terms. *Ecology* 16, 284–307.

Tansley, A. G. (1939). *The British Isles and Their Vegetation*. Cambridge University Press, Cambridge.

Thomas, C. D. (2017). *Inheritors of the Earth*. Allen Lane, London.

Thomas, C. D. & Palmer, G. (2015). Non-native plants add to the British flora without negative consequences for native diversity. *Proceedings of the National Academy of Sciences of the USA*, 112, 4387–4392.

Thomas, J. & Lewington, R. (2010). *The Butterflies of Britain and Ireland*, revised edition. British Wildlife Publishing, Gillingham.

Thomas, M. V., Malhi, Y., Fenn, K. M. *et al.* (2011). Carbon dioxide fluxes over an ancient broadleaved deciduous woodland in southern England. *Biogeosciences* 8, 1595–1613.

Thorne, K. S. and Blandford, R. D. (2017). *Modern Classical Physics*. Princeton University Press, Princeton.

Tipping, R. (1993). A detailed early postglacial (Flandrian) pollen diagram from Cwm Idwal, North Wales. *New Phytologist* 125, 175–191.

Tompkins, D. M., White, A. R. & Boots, M. (2003). Ecological replacement of native red squirrels by invasive greys driven by disease. *Ecology Letters* 6, 189–196.

Toogood, M., Waterton, C. & Heim, W. (2020). Women scientists and the Freshwater Biological Association, 1229–1950. *Archives of Natural History* 47, 16–28.

Tyrrell, T. (2013). *On Gaia*. Princeton University Press, Princeton.

Valverde, T. & Silvertown, J. (1997). A metapopulation model for *Primula vulgaris*, a temperate forest understory herb. *Journal of Ecology* 85, 193–210.

Van Damme, R., Wilson, R. S., Vanhooydonck, B. & Aerts, P. (2002). Performance constraints in decathletes. *Nature* 415, 755–756.

van der Heijden, M. G. A. & Horton, T. R. (2009). Socialism in soil? The importance of mycorrhizal fungal networks for facilitation in natural systems. *Journal of Ecology* 97, 1139–1150.

van der Heijden, M. G. A., Martin, F. M., Selosse, M.-A. & Sanders, I. R. (2015). Mycorrhizal ecology and evolution: the past, the present, and the future. *New Phytologist* 205, 1406–1423.

Vannier, N., Mony, C., Bittebiere, A.-K. *et al.* (2019). Clonal plants as meta-holobionts. *mSystems* 4, e00213-18. doi: 10.1128/ mSystems.00213-18.

Vitousek, P. (2004). *Nutrient Cycling and Limitation: Hawai'i as a Model System*. Princeton University Press, Princeton.

Vlk, L., Tedersoo, L., Antl, T. *et al.* (2020). Alien ectomycorrhizal plants differ in their ability to interact with co-introduced and native ectomycorrhizal fungi in novel sites. *ISME Journal* 14, 2336–2346.

Volk, T. (1998). *Gaia's Body: Towards a Physiology of Earth*. Copernicus, New York.

Walker, D. (1970). Direction and rate in some British post-glacial hydroseres. In D. Walker & R. G. West (eds.), *Studies in the Vegetational History of the British Isles*. Cambridge University Press, Cambridge, pp. 117–139.

Wall, T. & Wall, G. (eds.) (2019). *Rostherne Mere: Aspects of a Wetland Nature Reserve*. Tom Wall, Shropshire.

Waltham, D. (2014). *Lucky Planet*. Icon Books, London.

Ward, D., Kirkman, K. & Tsvuura, Z. (2017). An African grassland responds similarly to long-term fertilization to the Park Grass experiment. *PLoS ONE* 12(5), e0177208.

Warming, E. (1909). *Oecology of Plants*. Clarendon Press, Oxford.

Wetton, J. H., Carter, R. E., Parkin, D. T. & Walters, D. (1987). Demographic study of a wild house sparrow population by DNA fingerprinting. *Nature* 327, 147–149.

White, G. (1789). *The Natural History and Antiquities of Selborne*. B. White and Son, London.

Whittaker, R. H. (1969) New concepts of kingdoms of organisms. *Science* 163, 150–160.

Whittles, L. K. & Didelot, X. (2016). Epidemiological analysis of the Eyam plague outbreak of 1665–1666. *Proceedings of the Royal Society B* 283, 20160618.

Whitton, B. A. & Potts, M. (2012). Introduction to the cyanobacteria. In B. A. Whitton (ed.), *Ecology of Cyanobacteria II*. Springer, Dordrecht, pp. 1–13.

Wilkinson, D. M. (1998). Mycorrhizal fungi and Quaternary plant migrations. *Global Ecology and Biogeography Letters* 7, 137–140.

Wilkinson, D. M. (1999a). Is Gaia conventional ecology? *Oikos* 84, 533–536.

Wilkinson, D. M. (1999b). The disturbing history of intermediate disturbance. *Oikos* 84, 145–147.

Wilkinson, D. M. (2001). At cross purposes. *Nature* 412, 485.

Wilkinson, D. M. (2002). Ecology before ecology: biogeography and ecology in Lyell's 'Principles'. *Journal of Biogeography* 29, 1109–1115.

Wilkinson, D. M. (2004). The parable of Green Mountain: Ascension Island, ecosystem construction and ecological fitting. *Journal of Biogeography* 31, 1–4.

Wilkinson, D. M. (2006). *Fundamental Processes in Ecology: an Earth Systems Approach*. Oxford University Press, Oxford.

Wilkinson, D. M. (2012). Paleontology and ecology: their common origins and later split. In J. Louys (ed.), *Paleontology in Ecology and Conservation*. Springer, Berlin, pp. 9–22.

Wilkinson, D. M. (2018). What is a lichen? Changing ideas on the lichen symbiosis. *British Wildlife* 29, 351–357.

Wilkinson, D. M. & Sherratt, T. N. (2001). Horizontally acquired mutualisms, an unsolved problem in ecology? *Oikos* 92, 377–384.

Wilkinson, D. M. & Sherratt, T. N. (2016). Why is the world green? The interactions of top-down and bottom-up processes in terrestrial vegetation ecology. *Plant Ecology and Diversity* 9, 127–140.

Wilkinson, D. M., Creevy, A. L. & Valentine, J. (2012a). The past, present and future of soil protist ecology. *Acta Protozoologica* 51, 189–199.

Wilkinson, D. M., Koumoutsaris, S., Mitchell, E. A. D. & Bey, I. (2012b). Modelling the effect of size on the aerial dispersal of microorganisms. *Journal of Biogeography* 39, 89–97.

Wilkinson, D. M., Creevy, A. L. & Fisher, J. (2017a). Into the great unknown: the microbial diversity of a nature reserve. *British Wildlife* 29, 3–8.

Wilkinson, D. M., Lovas-Kiss, A., Callaghan, D. A. & Green, A. J. (2017b). Endozoochory of large bryophyte fragments by waterbirds. *Cryptogamie Bryology* 38, 223–228.

Williams, H. T. P. & Lenton, T. M. (2007). The Flask model: emergence of nutrient-recycling microbial ecosystems and their disruption by environment-altering 'rebel' organisms. *Oikos* 116, 1087–1105.

Willson, M. F. & Armesto, J. J. (2006). Is natural history really dead? Towards the rebirth of natural history. *Revista Chilena de Historia Natural* 79, 279–283.

Willson, M. F. & Burley, N. (1983). *Mate Choice in Plants: Tactics, Mechanisms, and Consequences*. Princeton University Press, Princeton.

Wilson, E. O. (1992). *The Diversity of Life*. Belknap Press, Harvard.

Wilson, J. B., Peet, R. K., Dengler, J. & Pärtel, M. (2012). Plant species richness: the world records. *Journal of Vegetation Science* 23, 796–802.

Wilson, J. B., Agnew, A. D. Q. & Roxburgh, S. H. (2019). *The Nature of Plant Communities*. Cambridge University Press, Cambridge.

Wilson, R. (2018). *Euler's Pioneering Equation*. Oxford University Press, Oxford.

Wohlleben, P. (2016). *The Hidden Life of Trees: What They Feel, How They Communicate.* William Collins, London.

Wood, K. A., Stillman, R. A., Daunt, F. and O'Hare, M. T. (2014). Chalk streams and grazing mute swans. *British Wildlife* 25, 171–176.

Woolway, R. I., Jones, I. D., Feuchtmayr, H. & Maberly, S. C. (2015). A comparison of the diel variability in epilimnetic temperature for five lakes in the English Lake District. *Inland Waters* 5, 139–154.

Wulf, A. (2015). *The Invention of Nature: the Adventures of Alexander von Humboldt, the Lost Hero of Science.* John Murray, London.

Yahr, R., Coppins, B. J. & Ellis, C. J. (2011). Preserved epiphytes as an archaeological resource in pre-industrial vernacular buildings. *Journal of Archaeological Science* 38, 1191–1198.

Yalden, D. W. & Albarella, U. (2009). *The History of British Birds.* Oxford University Press, Oxford.

Zambell, C. B., Adams, J. M., Gorring, M. L., & Schwartzman, D. W. (2012). Effect of lichen colonization on chemical weathering of hornblende granite as estimated by aqueous elemental flux. *Chemical Geology* 291, 166–174.

Zhalnina, K., Dias, R., De Quadros, P. D. *et al.* (2015). Soil pH determines microbial diversity and composition in the Park Grass Experiment. *Microbial Ecology* 69, 395–406.

Index

SPECIES INDEX

Note: page numbers in **bold** refer to information contained in captions and map locations. Page numbers in *italics* refer to information contained in tables.

GENERAL INDEX

Note: page numbers in **bold** refer to information contained in captions and map locations. Page numbers in *italics* refer to information contained in tables. Scientists mentioned in the main text are indexed, but not those only cited in parentheses as a reference.

The New Naturalist Library

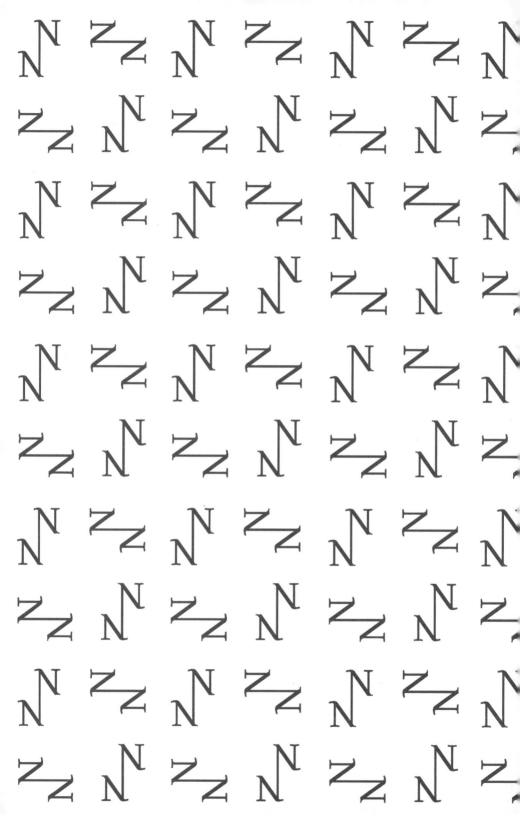